U0221032

回归与重生

城市双修探索与实践

风景融入日常生活

周欣萌　谢晓英　张琦　王欣·著

化学工业出版社
·北京·

图书在版编目（CIP）数据

回归与重生：城市双修探索与实践 / 周欣萌等著．
—北京：化学工业出版社，2024.3
（风景融入日常生活）
ISBN 978-7-122-44646-6

Ⅰ．①回… Ⅱ．①周… Ⅲ．①城市建设—研究
Ⅳ．①TU984

中国国家版本馆 CIP 数据核字（2024）第 000922 号

本书翻译（中译英）
Wang Yile（澳大利亚）　Daniel Lenk（英国）

审图号：GS 京（2024）0145 号

责任编辑：林　俐　　装帧设计：筑匠文化
责任校对：刘　一

出版发行：化学工业出版社（北京市东城区青年湖南街 13 号　邮政编码 100011）
印　　装：北京宝隆世纪印刷有限公司
710mm×1000mm　1/12　印张 14½　字数 250 千字　2024 年 3 月北京第 1 版第 1 次印刷

购书咨询：010-64518888　　售后服务：010-64518899
网　　址：http://www.cip.com.cn
凡购买本书，如有缺损质量问题，本社销售中心负责调换。

定　　价：98.00 元

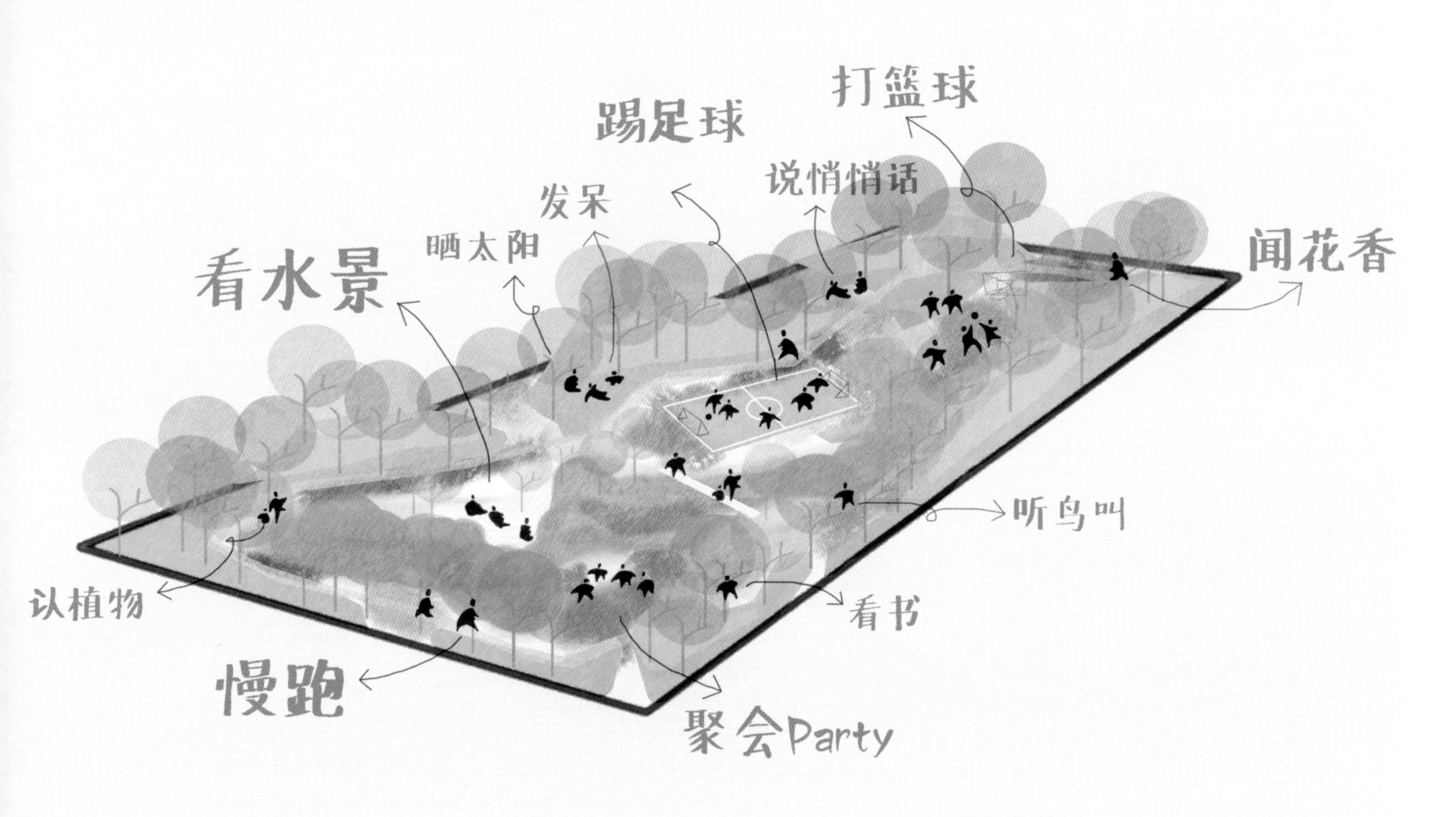

风景融入日常生活
Enjoy Dailylife

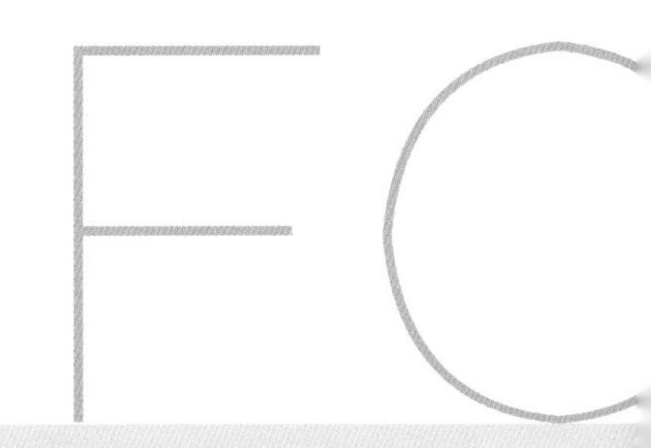

为无界景观设计团队的作品所作序

View Unlimited Design Team Project Preface

建筑、景观设计之为专业，不但有其功能意义，更关系一城一地的外部形象。在社会经济快速发展的时期，无疑处于“前沿”。景观设计的成果一旦落地，即成为项目所在地的日常现实，人们生活、呼吸其间，直至习为常态，日用而不自知。无论设计成果存在时间久暂，一种专业活动如此深地进入人们的生活，足以令人称羡。这一专业的“改变中国”，实实在在，不仅诉诸感官（不唯视觉），还影响人们的心态乃至生存状态。纵然有一天设计成果被时间抹去，其印在地面上尤其人们记忆中的痕迹，也是难以消失的吧。

景观设计与建筑设计，均为重塑空间的艺术：由物理空间到人文空间。景观设计师以大地为画布。无界景观设计团队的“绘画”作品散落于祖国大江南北以至不同的国家。团队一以贯之，从唐山凤凰山公园复建，到北京大栅栏杨梅竹斜街的改造，从北京城市副中心行政办公区核心区，到埃塞俄比亚首都亚的斯亚贝巴的友谊广场，均力求因地制宜、就地取材，低能耗，对原有的自然与人文环境少扰动，使人工与自然有机对接，满足多方面的功能需求。

团队在恪守专业人员的工作伦理的同时，不忘尽社会责任。团队的主要贡献在城市建设，却对“乡建”怀有志愿者般的热情；在城市作画之余，更将笔触伸向乡村。近年来“历史文化街区”“特色小镇”的建设一哄而起，以改造、升级为名，造成“千村一面”的不可逆的破坏，这样的现象亟待规范。优先考虑在地居民的需求，而非将他们的环境

Beyond the principle of functionalism, architecture and landscape design are directly related to the external image of cities and landmarks. In this period of rapid socio-economic development, it has taken on an increasingly prominent role. Once a landscape design is implemented, the project site becomes the daily reality for the surrounding inhabitants, who live and breathe it until they become accustomed to this new reality and over time, it naturally becomes a part of their everyday life. For as long as landscape design has existed as a profession, it has been admired for its ability to embed itself so deeply in people's lives. The transformational element of our profession is real — not only in terms of sensory perception, but also in terms of people's mindsets and state of being. Even if the accomplishments of design are erased over time, surely its remaining traces, ingrained in people's memories, are difficult to erase?

Landscape design and architectural design both focus on the art of reshaping space, from physical space to human space. The landscape architect uses the earth as their canvas, and the “brushstrokes” of View Unlimited’s works are scattered across northern and southern China, as well as various other countries. The View Unlimited team has been consistent in its approach, from the restoration of Phoenix Mountain Park in Tangshan to the renovation of Yangmeizhu Xiejie in Dashilan, Beijing, and from the administrative office area of the Beijing sub-center to Friendship Park in the Ethiopian capital of Adis Ababa. In each case, the team strove to leverage local conditions and materials, while emphasizing reduced energy consumption and minimal disturbance to the original natural and human environment, to create an organic interface between nature and man-made structures that meets a variety of functional needs.

The View Unlimited team has never forgotten its social responsibility, while adhering to their professional ethics. The team's main contributions fall within the scope of urban construction, yet they maintain a volunteer-like passion for rural development; while the team’s creations take form in the city, their bushstrokes extend to the countryside. Although in recent years, the construction of historical and cultural districts and "specialty towns" has been hastened in the name of transformation and modernization, it has caused irreversible damage to countless villages.This phenomenon needs to be regulated. Prioritizing the needs of local residents, rather than forcing them to sacrifice their environment and livelihood to become part of an exhibit that boosts the local economy — this is how landscape transformation can be

REWORD

连同生活作为展品，强迫其为拉动地方经济做牺牲——景观改造要这样才能永续。

中国的发展为景观设计专业提供了机遇。无界设计团队有幸参与这一进程，年轻设计师积蓄的能量得以释放。尽管负面的干预力量在所难免，人们环境意识以至社会审美风尚的进步，毕竟是更强大的力量。上述变化，由无界景观设计团队历年的项目即不难窥见。

于“隐身”——即不炫技、不刻意打造设计者自身形象——之外，这个团队贯穿始终的理念，更有“共享”。唐山凤凰山公园设计建造方便市民穿行的步道，赣西南夏木塘项目构建供乡民交流的公共空间，埃塞俄比亚友谊公园满足由官方仪式到民众日常娱乐的诸种需求，使广场成为便于人们交流融合的平台，他们的项目无不以社会效益为重要考量。

无界设计团队首先着眼的是项目所在地民众的日常感受；不取一时亮眼，而求可持续、可再生、可不断更新换代。在我看来，低调的美或更能持久。近年来追求“博眼球”，亮丽过后顿成鸡肋的项目比比皆是。回应民众需求，具备自我更新的能力，维持景观的持久活力，是对治此病的药方。无界设计团队还将改善当地民众的生存状况纳入优先考量，确保公共财政取之于民，用之于民，切实提升在地居民的幸福感。经历近几十年一波波的造景运动，团队的上述追求倍加可贵，值得阐发、揄扬。

长期实践中，团队形成了自己成熟的设计语言，如绿道、步道、步行网络的设计，又如对无机械健身，即“软性的、温和的、能够随时随地进行的健身”的推广。凡此，无不力求最大限度地利用空间，满足多方面的需

implemented sustainably.

China's development provides ample opportunities for landscape architects. The View Unlimited team was fortunate to be a part of this process, allowing for the pent-up energy of their young designers to be unleashed. Although the negative intervention force is inevitable, the progression in people's environmental consciousness and social aesthetics is a powerful and driving force. These changes can be easily seen in projects carried out by View Unlimited over the years.

In addition to playing the role of the "invisible hand" - that is, not to make grand statements in designs nor deliberately instill one's own image, the View Unlimited team emphasize the concept of sharing and exchange. Tangshan Phoenix Mountain Park's design and construction provides easy access for the public to navigate the mountain trail; the Xia Mu Tang landscape enhancement project in southwest Jiangxi Province features a public space that facilitates communication between the villagers; and Friendship Park in Ethiopia meets the needs of an assortment of people groups, ranging from official ceremonies to daily leisure, making the square a platform for various communities and social strata to come together.

The View Unlimited design team focuses first and foremost on the daily experience of the residents around the project site; not on brief flashes of glamor, but sustainability, regeneration, and continuous renewal. In my personal opinion, it is the beauty of being understated that has a greater permanence. In recent years, projects abounded that chased glamor, with little to no substance. The remedy to this condition is to respond to the needs of the people, embrace the ability to be flexible, and preserve the dynamism of the original landscape. The View Unlimited team also prioritizes improvements in local people's living conditions, to ensure that public funds are allocated appropriately and effectively, so as to increase the happiness of local residents. Considering the various movements in landscape design in recent decades, the team's above-mentioned pursuit is even more valuable and deserving of further exploration and advocacy.

Over time, the View Unlimited team has developed its own specific design language, which incorporates greenways, trails and walking networks, and the promotion of accessible fitness systems that can be used anytime and anywhere.

求。设计团队专业技术与人文层面并重，不但尊重自然，而且顺乎环境的历史脉络。对场地破拆材料进行再利用，既节约能源，又有效地保存场地记忆；以艺术化的地面、立面铺装，借助物料上的时间刻痕保存历史信息，都是团队行之有效的设计手法。砖石往往是既可视又可触的历史。镶嵌在地面、墙体的符号化的历史，以其物质形态嵌入了现实，不同时空的融汇借此种细节显现。用有年代感的砖、不同质料的石材，拼贴成图案承载记忆，这一设计手法在杨梅竹斜街改造、一尺大街修复、北京城市副中心行政办公区的设计与施工中，均有成功的运用。北京城市副中心行政办公区核心区步行道边镌刻北京新老胡同名，小广场铺设嵌入取自胡同的石材、木料，无不将老北京的文化密码不张扬地嵌入墙体、地面，只待行走者辨识。设计团队更注重利用现有的科技手段，实现“历史文化信息库的搭建，实现线上线下空间交叠”，既丰富了人们的空间体验，又使民众于娱乐休闲的同时摄取历史知识。凡此种种，随处可感用心之细，用情之深。

较之纸上的画作，实现在空间的绘制更难抵抗时间的侵蚀。景观设计作品势必经历一轮轮的更新、升级，也由此重生。原有设计中的亮点，有可能烛照、开启以后的设计思路。这也是设计团队为后续的“创作”预留空间的必要性及意义所在。

我一再说自己是书斋动物，因此对所有切切实实促进现状改善的努力都怀有敬意，对于景观设计专业不得不应对的诸种难题略知一二：“设计”不能止于图纸、文案，须将设想落实到施工现场，以至监督、干预用料的制作；具体实施中更要与相关各方磨合，做不得已的妥协。所有这些，岂是我这样的书生所能应对！作为行外人士，我欣赏的，毋宁说更是无界设计团队的理念，即如“安住”“乐活”，另如“微更新”“微改造”“微设计”“轻介入”“轻资产运营”“低成本开发”“小尺度下的微改造”，以至“如针灸一般

In all cases, the team strives to maximize the use of space and meet a diverse range of needs. The design team's technical expertise and user-friendly elements take on equal importance, not only in respect to the surrounding nature, but also the historical legacy of the environment itself. The utilization of a site's deteriorated materials not only conserves energy, but also effectively preserves the memory of the site, while artistic treatment to ground paving and facade cladding, combined with the preservation of historical information through dates engraved on materials have been carried out to great effect by the team. Masonry is often a visible and palpable part of history, and the symbolic history embedded in a site's paving and engraved on its walls is a reflection of reality in its material form. The convergence of different eras is revealed in such details, while the use of period-specific bricks and stones of different textures creates a collage of patterns that convey rich memories. This design technique has been successfully used in the renovation of Yangmeizhu Xiejie, the restoration of Yichi Dajie, and the design and construction of the administrative office area in Beijing's sub-center. The core area of the administrative office area in Beijing's sub-center features pedestrian paths engraved with the names of old and new Beijing hutongs, while small squares are paved with stone and wood taken directly from the hutongs in an effort to embed the cultural dna of old Beijing with no fanfare, but instead waiting to be discovered by visitors. The design team paid great attention to the use of existing technologies to achieve the construction of a "historical and cultural information base spanning both online and offline spaces", to not only enrich people's spatial experience, but also present historical knowledge in a refreshing and entertaining way. In all such cases, one can feel the attention to detail, and the rich emotions that are conveyed.

Such creations, which exist in our spatial environment, are more resistant to the erosion of time than a diagram concept. Landscape designs are bound to undergo multiple rounds of updates, improvements, and ultimately rebirth. Yet, the salient points of an original design have the potential to illuminate and spark future design concepts. This is why it is both necessary and meaningful for design teams to retain space for subsequent creativity.

Needless to say, I am a bookworm. As such, I have great respect towards any and all initiatives to promote meaningful improvements to the status quo, and I have come to understand the challenges that landscape design professionals are faced with — that design does not end in blueprints and reports, but rather, must be implemented in the construction site itself, and require supervision and even intervention in the production of materials used. Cooperation between all of the parties involved, and making necessary compromises, is paramount to the practical implementation of a project. Frankly, as a scholar, such considerations can be overwhelming! However, as a layman, I appreciate

微创与介入式的改造”——切中时弊，较之具体项目，或许更有推广的价值。对成效的估量或有不同，上述原则对于相关行业，值得一再重申。无界设计团队所实行的“微改造”，或许多少也因财力所限，但也不妨换一种思路：这也是在严格给定的条件下创造最大的效益。

无界设计团队一再扩大作业范围，由京城而外地而乡村而国外。项目对象有北京城市副中心，也有村落；项目性质有外地商业性质的，也有带有公益性质的京城旧城区改造，以及援外项目，无界设计团队无不从容应付，并因项目性质的变换、场域的转移、涉及范围的扩展而自我提升、完善，化蛹成蝶。尽管岁月轮转，时间推移，这一团队长时间地保持生机勃发，展现出不会枯竭的原创性与潜能。我有幸或多或少地见证了这一过程，领略了团队虽受困于现实条件但仍不放弃专业理想与职业操守的顽强，也由他们坚持的理念、思路获得滋养。

对于无界设计团队业绩的评估，既有空间的也有时间的尺度。无论怎样，在我看来，正是这种一点一滴的改变，一片区一角隅的重新塑造，影响着未来中国的面貌。纵然因形格势禁，一些富于创意的设计未能实现在地面上，也以设计图、文案的形式为行业提供借鉴。纸墨更寿于金石，为一个时代的行业状况留文献，岂不是有不可替代的价值?

业界将包括无界景观设计在内的一些设计团队的项目介绍称为“来自前线的报告”，甚得我心。在一轮轮的城市改造、“新农村”建设潮中，建筑、景观设计的确位于“前线”。关心未来中国样貌的人们，无疑希望继续收到这种“来自前线的报告”。

the philosophies the View Unlimited design team embrace, such as "live and let live", "micro-renewal", "micro-renovation", "micro-design", "gentle intervention", "asset-light operations",

"austere development", "small-scale micro-renovation", and even “targeted, minimally invasive and interventional renovation". Each of these concepts are relevant, and perhaps worthier of promotion than any one project. Valuations of the team's achievements may vary, but the above principles are worth reiterating again and again in both our industry and other related industries. The micro-renovations implemented by the View Unlimited design team have at times been constrained to a certain degree by financial resources, but from another perspective, the team has created the greatest possible benefits under strict limitations.

View Unlimited has repeatedly expanded the scope of its work, implementing projects in the capital, the suburbs, the countryside, and even abroad. The projects involved range from Beijing's sub-center to far-flung villages, while the nature of the projects range from overseas commercial projects, to public welfare projects such as the renovation of old Beijing, and even foreign aid initiatives. Regardless of the project in question, the View Unlimited design team operates with poise and confidence, and despite the fluid nature of the projects they undertake, changes to the site, or the increasing scope of projects, they continue to hone and perfect their approach. Despite the years that have passed, this group of people has remained active, and continue to demonstrate their originality and inexhaustible potential. I was fortunate enough to witness this process to a greater or lesser extent, to appreciate the tenacity of a team that refuses to give up on their professional ideals and ethics despite the realities of the industry, and to be spiritually nourished by the ideas and principles they continue to pursue.

View Unlimited's performance can be evaluated on both a spatial and temporal scale. In either case, I believe that the team's initiatives are defined by incremental changes, such as the reshaping of a corner or neighborhood. Yet over time, such changes will come to define the image of China's future. Although some creative designs cannot be realized due to prevailing limitations, they live on in the form of illustrations and writings, which the industry can continue to draw inspiration from. Wisdom holds that paper and ink are more enduring than gold and stone; would it not be of irreplaceable value to preserve such documentation of the state of the industry in our era?

The project presentations of design teams such as View Unlimited Landscape Design have been referred to by the industry as "reports from the front line", which I find rather enthralling. Indeed, in the next round of urban transformation defined by new rural development, architecture and landscape design are indeed on the "front line". For those who have an interest in what China will look like in the future, they will no doubt want to continue to receive such reports from the front.

前言
PREFACE

2017年，中国住房和城乡建设部提出将“城市双修”作为城市规划的指导性理念。城市双修包含城市修补和生态修复，城市修补指利用织补更新手段，完善城市功能，盘活城市资源，塑造城市特色；生态修复指修复城市破损的生态环境，改善环境质量。

“城市双修”倡导人与自然的和谐共生，是符合中国当前发展阶段的一种城市建设方式。中国的城市发展正处于转型阶段，其特点是从以GDP增长为中心的发展模式转向以人为本、同时注重规模和质量的新阶段。城镇化正在从数量增长转变为质的提升，发展正在从投资驱动转向创新驱动，建设方式正在从单纯的城市管理向更全面的城市治理模式转变。对城镇化质量的评价越来越强调个人的主观经验和感知。在这种情况下，城市建设者要转变城市发展方式，保护好环境，并提高居民的生活质量。城市变得比以往任何时候都更加智能。在此背景下，“城市双修”侧重于解决国家快速发展留下的创伤，其中包括不受控制的城市扩张导致的环境退化，粗放开发造成的资源浪费、城市功能不足，以及在以经济为中心的发展模式中出现的缺乏当地文化的同质城市景观。

“城市双修”是一种可持续的城市更新方法，是不断调整人与自然之间，以及人与社会之间关系的过程。随着我国新型城镇化战略的推进，城市发展正从激烈的城市竞争时期转向区域协调发展时期。作为新时代的指导思想，“城市双修”强调区域范围内更广泛的城市发展，通过最大化管辖空间或超越行政边界来促进不同规模的城市协调发展，促进产、城、人的进一步融合。这种方法还鼓励城乡一体化，形成从乡村到城镇再到大城市的

In 2017, the Ministry of Housing and Urban-Rural Development of the People's Republic of China proposed "dual urban regeneration" as a guiding concept for urban planning. Dual urban regeneration includes urban repair and ecological restoration. Urban repair involves utilizing innovative renewal strategies to enhance urban functionality, revitalize urban resources, and shape distinctive urban characteristics. On the other hand, ecological restoration focuses on repairing the damaged urban ecosystem and improving environmental quality.

Dual urban regeneration is an approach aligned with China's current stage of development, advocating for a harmonious coexistence between humans and nature in urban construction. China is in the midst of an urban transformation marked by a shift from a development model centered on GDP growth to a new phase that prioritizes people and emphasizes both scale and quality. Urbanization is transitioning from quantitative growth to qualitative improvement, development is shifting from being investment-driven to innovation-driven, and construction approaches have evolved from mere urban management to a more comprehensive model of urban governance. The evaluation of urbanization quality increasingly emphasizes the subjective experiences and perceptions of individuals. In this context, dual urban regeneration focuses on addressing the wounds left by the rapid development of the country, These include issues such as environmental degradation resulting from uncontrolled urban sprawl, wasteful resource consumption due to extensive development, insufficient urban functions, and the emergence of homogeneous cityscapes devoid of local culture in an economy-centric development model.

Dual urban regeneration represents a sustainable approach to urban renewal that requires ongoing adjustments in the relationships between people and nature, as well as between people and society. With the progress of a new urbanization strategy, urban development in China is shifting from a period of intense urban competition to one of coordinated regional development. Serving as a guiding ideology for this new era, dual urban regeneration places emphasis on broader urban development at a regional scale and the increased integration of industry, cities, and people by maximizing jurisdictional spaces or transcending administrative boundaries to promote the coordinated development of cities at various scales. This approach also encourages the integration of urban and rural areas, creating

连续统一体，利用不同的人口规模促进创新发展和成本优势，从而为更具创新性的城市发展和增强城市竞争力提供空间。此外，“城市双修”的城市更新方法能降低生活成本，同时进一步区域的基础设施建设和公共服务供给，更符合大量中低收入个人的住房和就业需要。

“城市双修”是推进绿色可持续城市发展、促进生态文明建设、促进全球生态安全的途径。人们在经历了疫情、气候变化和其他各类全球挑战之后，对更健康的城市生活环境的需求不断增长。根据联合国人类住区规划署《2022年世界城市报告》，城市化仍然是21世纪的发展趋势，城市的未来与人类的未来交织在一起。韧性发展是城市未来的核心。中国的城市化进程远未完成，城市居民还面临着与经济增长、生活水平和生计相关的各种严峻挑战。“城市双修”通过加强经济、社会、环境和制度弹性来应对这些挑战，以塑造高质量的城市空间，建立强大的产业平台，并挖掘新的经济增长动力。“城市双修”促进了绿色低碳生产方式、生活方式和城市建设运营模式的采用，使城市做好抵御未来各种冲击的准备，更好地实现可持续发展。

景观设计是最直接处理人与自然关系的专业，本书收录了中国城市建设研究院无界景观工作室近二十年的若干项目实践，着重探讨了在中国经济转型时期，景观设计利用专业技术进行生态恢复和城市修复，在重塑城市、协调区域发展和增强韧性方面所发挥的关键作用。在此过程中，景观设计释放了全部潜力和价值，为在城市和乡村景观中创造整体而持久的美丽做出了重要的贡献。

a continuum from villages to towns to major cities that leverages different population sizes to facilitate innovative development and cost advantages, thereby providing space for more innovative urban development and enhanced urban competitiveness. Moreover, this approach aligns more closely with the housing and employment needs of a significant number of middle and low-income individuals, taking into account their cost of living, while further promoting infrastructure development and the provision of public services in such areas.

Dual urban regeneration represents a pathway towards advancing green and sustainable urban development, promoting an ecological civilization, and contributing to global ecological security. In the wake of the COVID-19 pandemic, climate change, and other global challenges, there is a growing demand for healthier urban living environments. According to the United Nations Human Settlements Programme's *2022 World Cities Report*, urbanization remains a prevailing trend in the 21st century, and the future of cities is intertwined with the future of humanity. Resilient development lies at the core of the urban future. China's urbanization process is far from complete, and it faces formidable challenges related to economic growth, living standards, and livelihoods. Dual urban regeneration addresses these challenges by strengthening economic, social, environmental, and institutional resilience in order to shape high-quality urban spaces, establish robust industrial platforms, and tap into new economic growth drivers. This approach promotes the adoption of green and low-carbon production methods, lifestyles, and urban construction and operation models. By preparing cities to withstand various future shocks, dual urban regeneration facilitates more effective strides towards sustainable development.

Landscape design is unique in its direct engagement with the intricate relationships between humans and nature. This book serves as a documentation of View Unlimited Studio's landscape design endeavors spanning the past two decades. Its primary focus is to delve into how, amid China's profound economic transformation, landscape design has harnessed professional techniques for ecological restoration and urban repair, playing a pivotal role in reshaping cities, harmonizing regional development, and bolstering resilience. In doing so, landscape design unlocks its full potential and value, ultimately making significant contributions to the creation of a holistic and enduring aesthetic across both urban and rural landscapes.

在促进城市转型方面，我们通过探寻因地、因时、因人而异的场地问题的解决方案，恢复人与自然的身心联系，塑造健康的人居环境。这一探索始于2005年"唐山凤凰山公园改造"项目，景观设计从修复被破坏的山体和植被入手，强化了凤凰山这一城市历史地标的认知度；同时，建立穿行网络（路网），模糊城市与公园的边界，将自然更好地融入市民的日常生活；构建户外交往场所，衍生出多种休闲活动，以一种"快乐、健康、环保"的生活方式，缓解、释放压力，促进社会和谐，市民成为城市新形象的塑造者和传递者。在后来的"鄂尔多斯20+10中小企业园区景观设计"和"北京中信金陵酒店景观设计"等项目中，我们也延续了这一设计思路：基于对场地的理解赋予场地性格，创造人与环境之间的最佳关系，建设具有引导性、参与性和吸引力的户外活动空间，将人们引导向更健康的生活方式。

在协调区域发展方面，我们通过绿道、郊野公园体系建设，形成山水林田湖生态系统和景观风貌连续体，促进城乡融合发展。在"漳州郊野公园体系建设和环境整治"项目中，我们将城乡连续体作为研究对象，从解决具体生态问题入手，采取系统的思维方式，将山、水、林、田、湖、城作为一个有机的生命体，保护乡土风貌与自然生态格局，建立城乡融合的一体化空间结构，提升城镇带动乡村的集聚效应和辐射带动作用，实现城乡统筹协调发展。在"厦门市海沧区绿道统筹建设"项目中，我们将绿道和城市公共景观作为绿色基础设施，保护生态环境，控制城市的无序扩张，建立了一个能够整合各种资源，使城市一体化发展的平台。景观设计在解决生态问题的同时延续地域文化，带动文教、体育、旅游等产业发展，引导公众参与城市建设，培养市民健康积极的生活方式。

In our pursuit of urban transformation, we have explored location, period, and people-specific solutions aimed at restoring the physical and mental connection between people and nature while nurturing a healthy living environment. This journey began with the Tangshan Phoenix Mountain Park Renovation in 2005, where we undertook the restoration of damaged terrain and vegetation to reinforce the identity of this historic landmark. An interweaving network was established to blur the demarcations between the city and the park, seamlessly integrating nature into the fabric of daily life. The creation of outdoor gathering spaces has cultivated a diverse range of activities, promoting a "joyful, healthy, and eco-friendly" lifestyle. This holistic approach not only alleviated stress but also promoted social harmony, effectively transforming citizens into shapers and conveyors of the city's revitalized image. Subsequent projects, such as the Ordos 20+10 Small and Medium Enterprise Park and Beijing CITIC Jinling Hotel, continued to embrace this design philosophy. By delving deeper into the unique attributes of each site, our aim was to cultivate the perfect synergy between individuals and their surroundings, while encouraging the adoption of healthier lifestyles through the development of engaging outdoor activity spaces.

In the realm of coordinating regional development, our primary focus has revolved around the establishment of greenways and suburban park systems. This strategic approach has given rise to a continuous ecological network and landscape continuum that encompasses mountains, water bodies, forests, fields, lakes and urban areas. Such an integrated approach serves as a catalyst for the harmonization of urban and rural areas. In the Zhangzhou Suburban Park System and Environmental Revitalization Project, we systematically addressed ecological issues by considering mountains, water, forests, fields, lakes, and cities as interconnected elements of a unified organism. This approach sought to preserve local culture and native landscapes and ecologies while structuring integrated urban-rural spaces, enabling cities to stimulate rural development through clustering and diffusion effects to achieve a coordinated model of urban-rural development. In the Xiamen Haicang District Greenway Project, we recognized greenways and public landscapes as essential components of green infrastructure. They played a pivotal role in preserving ecologies and curbing urban sprawl, while establishing a platform capable of integrating resources to facilitate the holistic development of the city. By addressing ecological concerns while embracing regional culture, landscape design became a catalyst for various sectors, including education, sports, tourism, and more. This approach encouraged public participation in urban development and fostering the cultivation of healthy, active lifestyles.

在加强韧性建设方面，我们通过景观统筹、多专业协作，进行绿色基础设施、海绵城市等的建设，高效利用土地资源，探索高质量发展、低自然消耗的城市建设模式。在“重庆两江新区悦来新城后河环境综合整治工程”和“湖南益阳市梓山湖片区规划设计”项目中，我们通过景观设计统筹流域综合整治、海绵城市、生态景观、市政综合管网、智能监控监测等内容，建设绿色市政基础设施，探索新城建设模式，高效利用土地资源。

二十大报告提出，现阶段我国社会的主要矛盾是“人民日益增长的美好生活需要和不平衡不充分的发展之间的矛盾”。“城市双修”理念提倡一种更加细致和人性化的城市发展模式，更加强调尊重基层社会需求和创造振兴机会。在景观设计工作中，我们将继续探索有效的策略，以协调个人与环境之间以及个人与社会之间的关系。我们的目标是实现环境保护与社会经济发展之间的平衡，并为全球尤其是广大发展中国家的城市化和可持续发展提供可借鉴的解决方案。

中国城市建设研究院无界景观工作室

2023年10月

In our efforts to bolster environmental resilience, we have actively embraced landscape design that incorporates green infrastructure and sponge city construction through interdisciplinary collaboration. This strategy optimizes land resources and explores urban development models that prioritize high-quality growth while minimizing the consumption of natural resources. Notable examples of our work in this domain include the Chongqing Liangjiang New Area Yuelai New City Houhe Environmental Remediation Project and the Hunan Yiyang City Zishan Lake Scenic Area Urban Design Project. In these projects, landscape design assumes a central role in facilitating a unified approach to basin remediation, sponge city initiatives, ecological landscapes, municipal networks, and intelligent monitoring and surveillance systems to construct green municipal infrastructure and pioneer innovative urban development models, all while maximizing the efficient use of land resources.

The 20th Party Congress Report underscores that the principal contradiction of Chinese society lies in reconciling the increasing aspirations of its people for a better life with the existing imbalances and insufficiencies in development. The concept of dual urban regeneration promotes a more nuanced and humane model of urban development that places greater emphasis on respecting grassroots social demands and fostering opportunities for revitalization. In our endeavors, we will continue to explore effective strategies to harmonize the relationships between individuals and the environment, as well as between individuals and society. Our goal is to achieve an equilibrium between environmental protection and socio-economic development and provide solutions a valuable reference for urbanization and sustainable development on a global scale, especially for developing countries.

View Unlimited Landscape Architecture Studio, China Academy of Urban Planning and Design

October 2023

目录 CONTENTS

生态修复
技术应用

APPLICATION OF ECOLOGICAL REMEDIATION TECHNOLOGY

生态修复是在生态学原理指导下，以生物修复为基础，结合各种物理修复、化学修复以及工程技术措施，通过优化组合，使之达到最佳效果和最低耗费的一种综合的修复污染环境的方法。无界景观以“万物是互联的”系统思维为指导，将多种生态修复技术有机组合，提出适合当地自然和经济条件的环境保护和生态修复策略，实现人工和自然环境的连通和整体化，建立人与自然的多维链接，倡导绿色低碳的生产、生活方式，促进城市转型发展。

Ecological remediation is a comprehensive method of remedying polluted environments guided by the principles of ecology, with biological remediation as the foundation. It combines various physical remediation, chemical remediation, and engineering techniques in an optimized manner to achieve the best results with the lowest cost. Guided by the systemic thinking of "everything is interconnected," View Unlimited Landscape Architects Studio organically integrates multiple ecological remediation technologies and proposes environmental protection and ecological restoration strategies that are suitable for local natural and economic conditions. It aims to establish a connection and integration between artificial and natural environments, create multidimensional links between humans and nature, advocate for green and low-carbon production and lifestyle, and promote urban transformation and development.

01 鄂尔多斯 20+10 中小企业园区景观设计

Landscape Design of Ordos 20+10 Small and Medium Enterprise Park

- **项目地点：** 中国 鄂尔多斯
- **项目规模：** 1.53 平方千米
- **设计时间：** 2010 年

- **Project location:** Ordos, China
- **Project scale:** 1.53 square kilometers
- **Design period:** 2010

景观设计针对当地硬梁地貌与气候条件，将地形塑造与生态工程相结合，改造贫瘠的冲沟为植被茂盛、内容丰富的谷地公园，构建城市资源共享、景观与建筑“互联”的城市绿色公共空间。

The landscape design of the park focuses on the local ridge-and-valley landform and climatic conditions. It combines terrain shaping with ecological engineering to transform barren gullies into lush and diverse valley parks, building urban resources sharing, landscape and architecture“interconnection” of the city’s green public space.

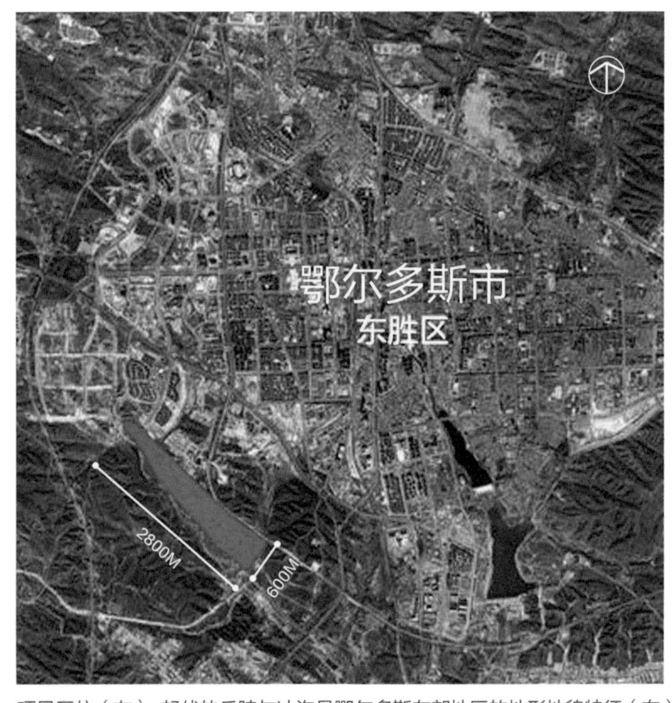

项目区位（左）；起伏的丘陵与冲沟是鄂尔多斯东部地区的地形地貌特征（右）

Project Location (left); The undulating hills and ravines are the topographic features of the eastern region of Ordos (right)

鄂尔多斯市地处内蒙古鄂尔多斯高原腹地，曾经水草丰美，但由于历史上气候变迁、战乱、放垦等诸多原因，生态环境逐渐恶化。改革开放以来，鄂尔多斯市高度重视生态文明建设，沙漠化和水土流失面积逐步减少，但因其薄弱的生态基底及半干旱的温带大陆性气候，生态环境依然敏感脆弱。

鄂尔多斯20+10中小企业园区位于鄂尔多斯市东胜区昆独龙沟北岸，主体是由20多名国内著名建筑师及10名国际建筑师设计的60栋各具特色的集群建筑。该园区不仅是鄂尔多斯未来的中小企业总部基地，同时还担负着商业、办公、休闲、旅游等多重功能，是城市的地标及综合体。

景观设计的核心理念是“云时代”，云时代即“云计算时代”，它包含的思想是把力量联合起来给其中的每一个成员使用，蕴涵了未来的发展趋势，即更加注重资源的链接、整合与共享。因此，我们首先统筹资源，修复生态环境创造宜人的室外场所，进而通过设计建立园区内部、园区与城市的多维连接，让任何人或物都可能成为资源网络的终端，突破时间与空间的限制使用这些资源。

统筹景观设计与生态环境的修复

景观设计注重与生态修复工程的统筹，针对鄂尔多斯市气候特点和硬梁地貌，采用地形塑造与生态集水、涵养水土、土壤改良相结合的手法，在现状可行条件中，运用生态技术手段，避免大量建设对生态环境的破坏。

结合植物自身对自然的改造和修复能力，重新构建植被体系与植物景观，利用植物群落改善环境小气候，将现状贫瘠的冲沟改造为植被茂盛、内容丰富的公园，提高环境舒适度，培养低碳环保、可持续的生活方式，使这里的人们生活在绿树成荫、空气湿润、优美宜人的“城市氧吧”“绿色家园”，增加居民的归属感、自豪感、幸福感，成为城市可持续发展的保证。

“互联”城市绿色公共空间

项目地块南侧为昆独龙沟滨河景观带，北侧为密集的城市居住区。地块内中小企业办公使用的60栋建筑单体位于五个抬升的高地上。景观设计针对建筑单体的不同特质，协调建筑之间的场地关系，通过地形改造加强起伏的地形特色，通过绿地和步行系统联系建筑与户外空间，构建景观与建筑交织互联的城市绿色公共空间，成为面向城市展开的优美画卷。

我们充分考虑项目与周边环境的有机共生关系，打开园区的边界，通过七个条状谷地公园构建视线通廊，规划可达的步行道路联系市民与公园，链接城市与河道。实现项目内土地的集约利用，使生产、生活与生态空间有机衔接，使景观与城市相互渗透并延伸。

从空间和时间上整合园区商业、文化、自然资源，增设小型的餐饮、购物服务设施和休憩设施，同时服务于园区和城市，随时随地将这些资源提供给需要者使用，并为使用者提供丰富多彩的游憩体验。项目以按需易扩展的方式提供丰富的空间和令人愉悦的场所，打造鄂尔多斯新时代的活力名片。

现状地形与植物
Existing Topography
and Vegetation

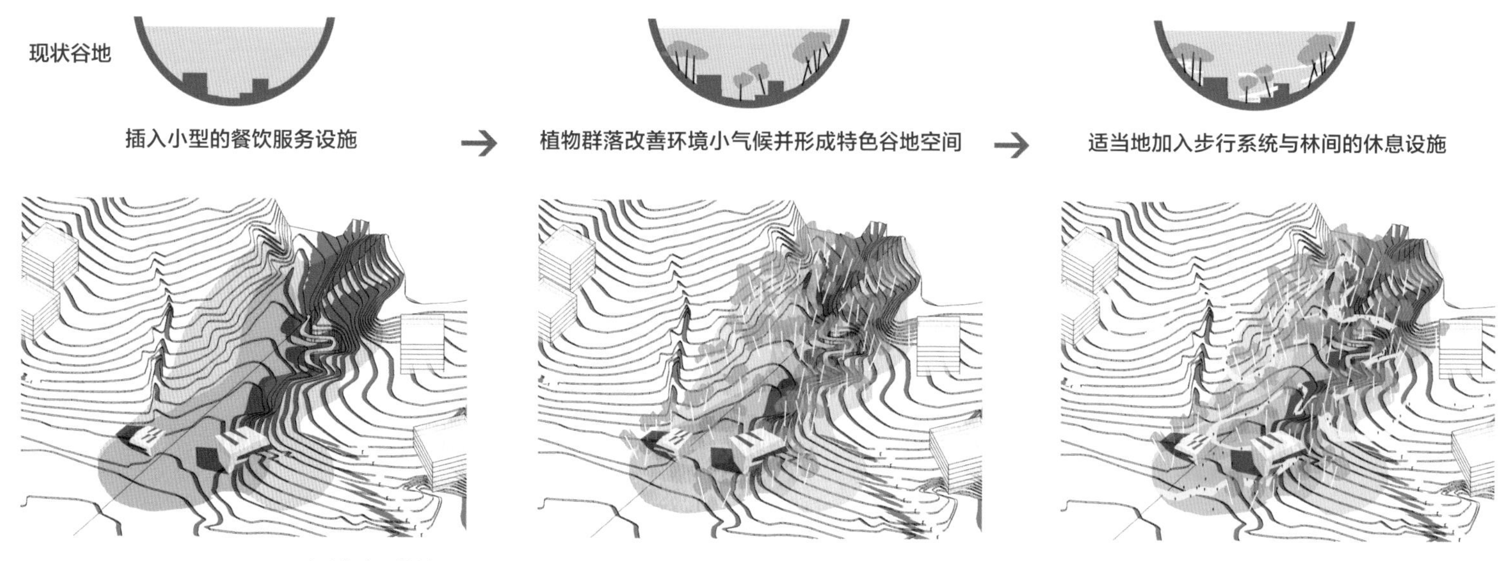

通过对现状地形改造利用，将环境恶劣的冲沟变成人们的“午间庭院”“室外会议室”“假日花园”

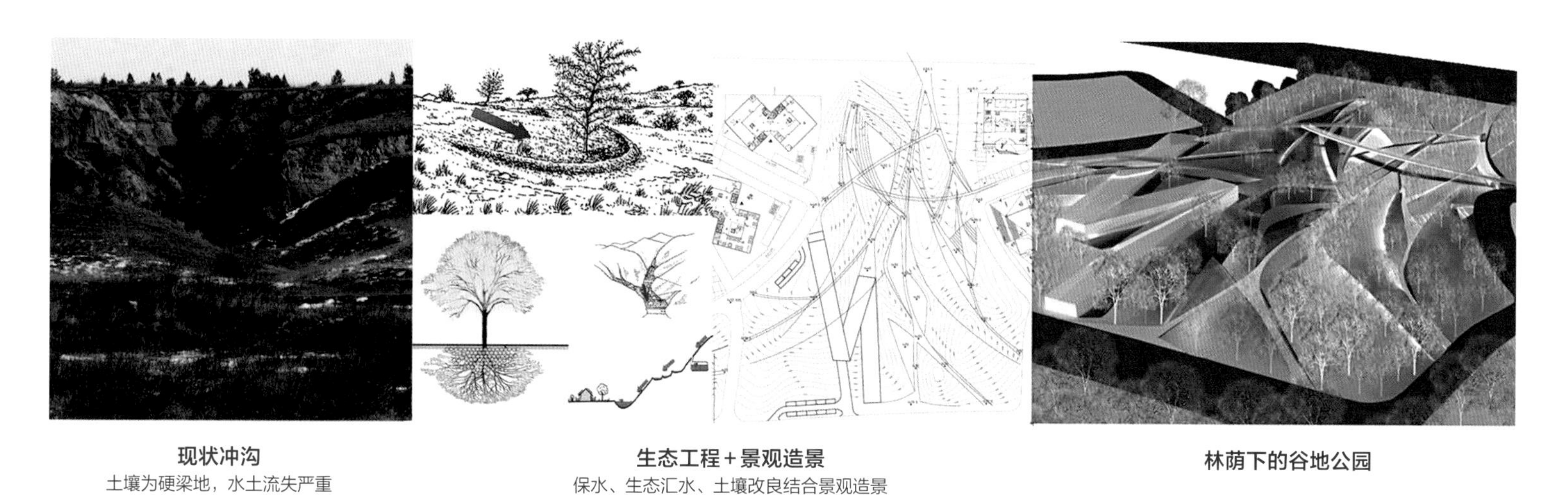

现状冲沟
土壤为硬梁地，水土流失严重

生态工程+景观造景
保水、生态汇水、土壤改良结合景观造景

林荫下的谷地公园

将生态集水工程与景观地形设计相结合，采用土壤改良剂，雨水收集蓄水池及植物生长保障措施，既防止水土流失，又形成景观空间
Combining ecological water collection projects with landscape terrain design, using soil improvers, rainwater collection reservoirs, and measures to ensure plant growth, not only prevents soil erosion but also creates a landscape space

排水工程与造景结合，结合冲沟地势设计可举办球赛、那达慕大会等活动的场地

Combining drainage engineering with landscaping based on the terrain of ravines, the design of the site can accommodate activities such as ball games and Nadam festivals

景观集水池创造良好的小气候
Landscape collection ponds create a favorable microclimate

项目基地南侧的山峦和昆独龙沟景观

景观与建筑设计共同加强原本起伏的地形特色

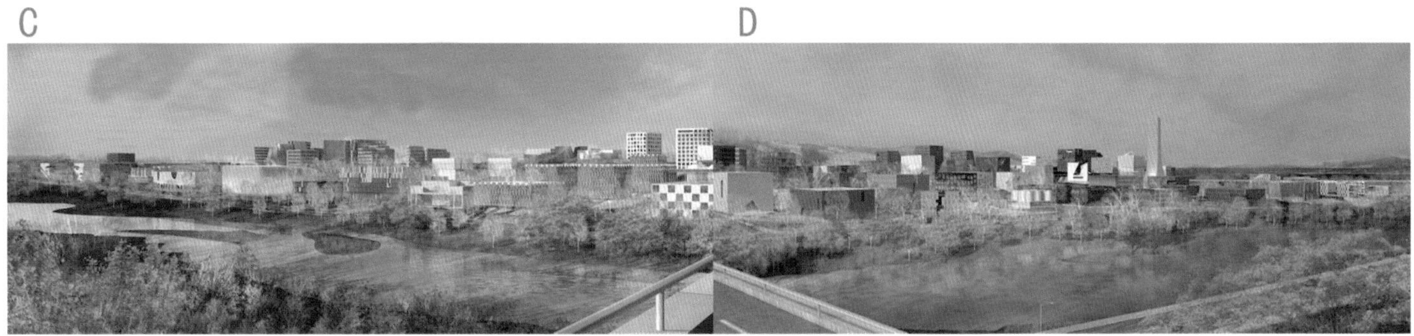

景观设计针对建筑单体的不同特质，通过地形改造加强起伏的地形特色
The landscape design enhances the undulating terrain features by transforming the terrain according to the different characteristics of individual buildings

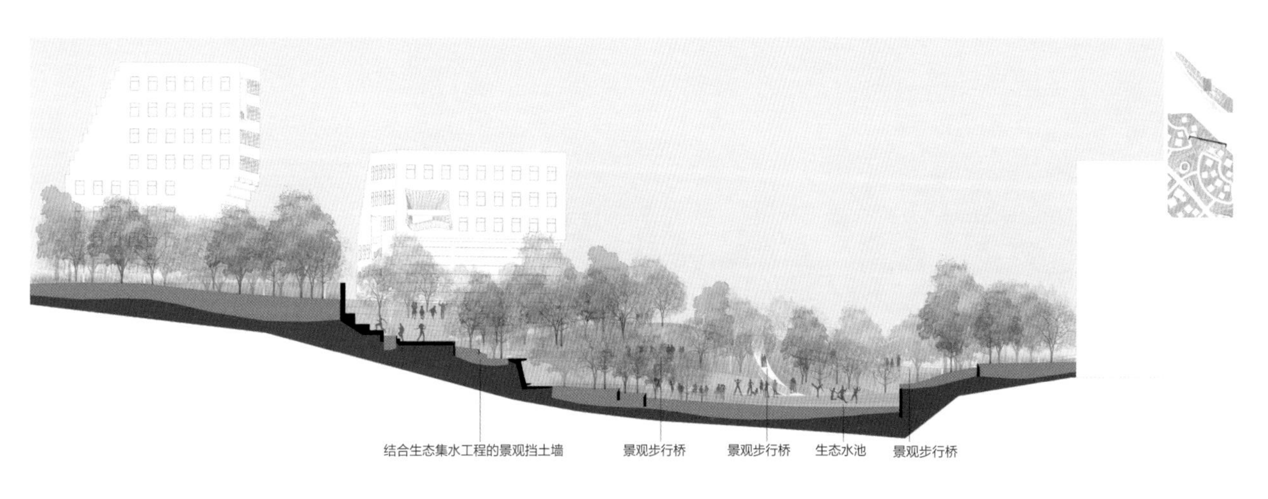
结合生态集水工程的景观挡土墙
景观步行桥
景观步行桥
生态水池
景观步行桥

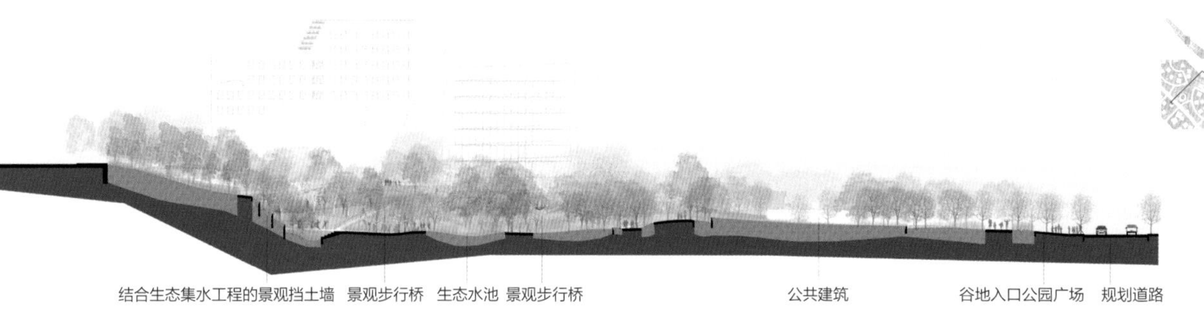
结合生态集水工程的景观挡土墙
景观步行桥
生态水池
景观步行桥
公共建筑
谷地入口公园广场
规划道路

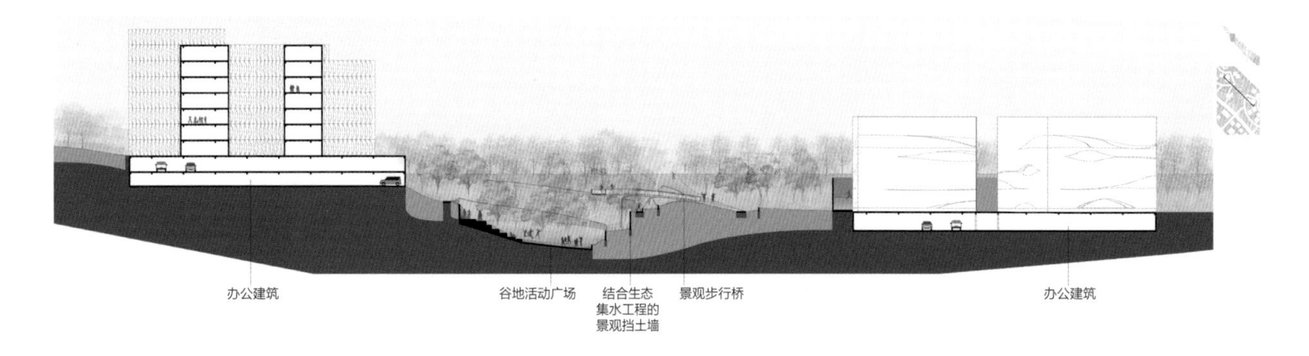
办公建筑
谷地活动广场
结合生态
集水工程的
景观挡土墙
景观步行桥
办公建筑

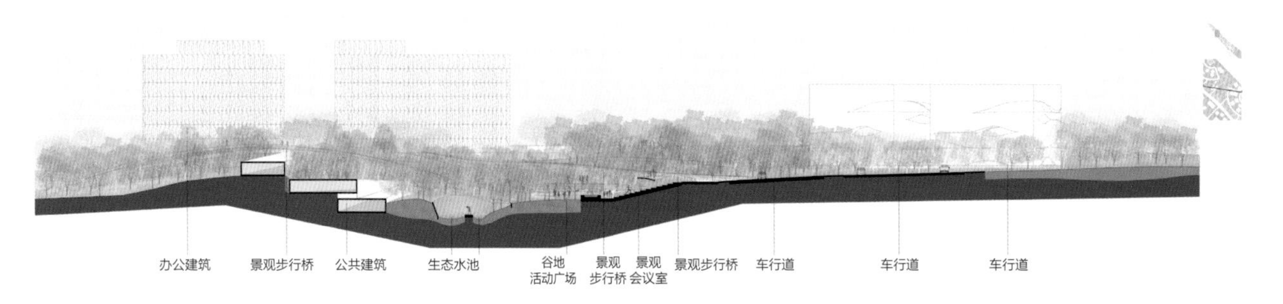
办公建筑
景观步行桥
公共建筑
生态水池
谷地
活动广场
景观
步行桥
景观
会议室
景观步行桥
车行道
车行道
车行道

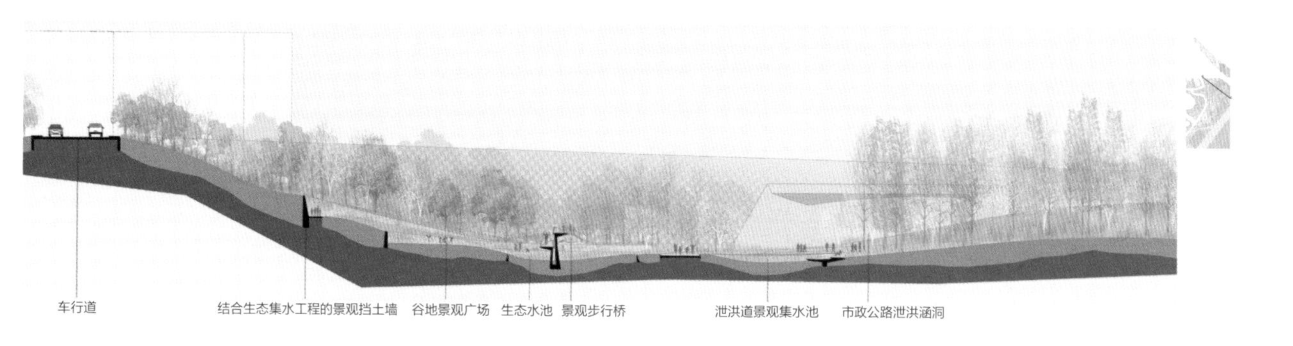

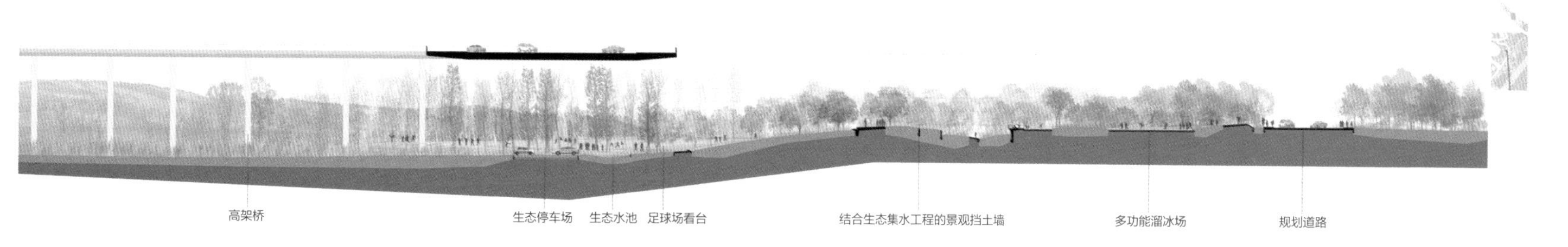

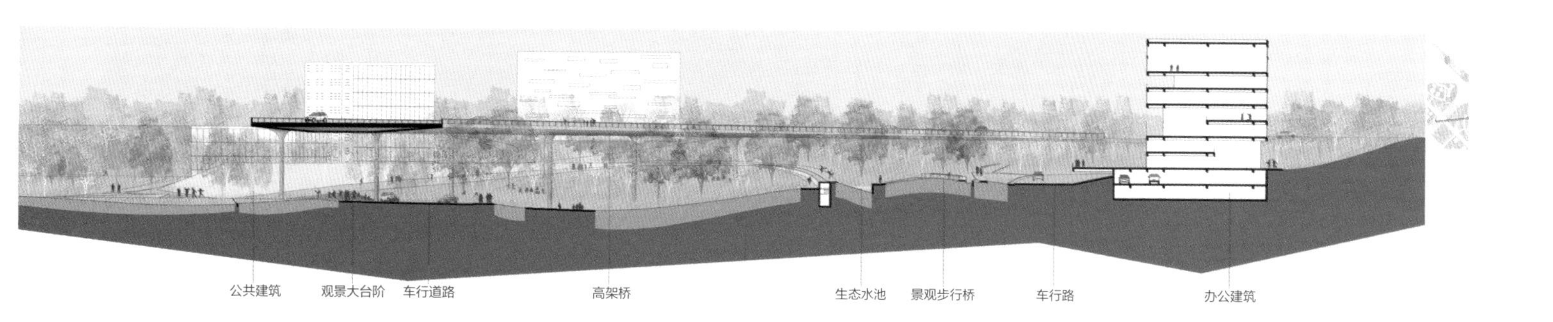

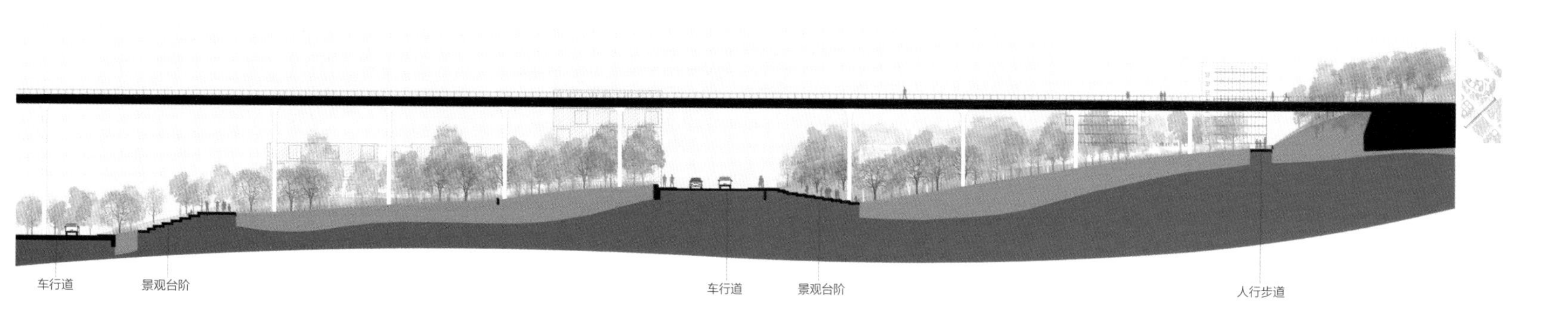

设计剖面图候 Design section drawing

a 树林公园轴：以树林景观为主，是天然氧吧
b 入口公园轴：面向入口道路的绿色休闲场所，给人们以良好的空间和视觉体验
c 谷地公园轴结合坡地、设施和林下广场的绿色山谷休闲区
d 桥下商业广场活力轴：高架桥与绿地交汇的休闲区，可结合场地特色举办艺术和创意活动
e 运动公园轴：开阔的绿地中设置公园式的运动场地，鼓励人们到户外健身

地块内因地势和功能要求分布7条特色景观轴，穿插在园区内，丰富了园区的景观空间

Within the lot, 7 characteristic landscape axes are distributed based on terrain and functional requirements, interspersed within the park, enriching the park's landscape space

0 物业管理
1 酒店与会议中心
2 餐饮娱乐与综合服务
3 餐饮与室内休闲
4 公寓、托儿所、商业、综合服务
5 商业综合服务
6 树林公园
7 景观谷公园
8 桥下活力广场
9 活力健身谷公园
10 高压走廊
公交车站

中心道路景观效果图
Rendering of the central road landscape effects

高架桥下，景观与公共建筑结合的活动空间效果图
Activity Space Rendering of Landscape and Integrated Public Building Under Elevated Bridge

02 北京中信金陵酒店景观设计

Landscape Design of Beijing CITIC Jinling Hotel

- **项目地点：** 中国 北京
- **项目规模：** 0.1063 平方千米
- **设计时间：** 2011-2015 年
- **施工时间：** 2013-2017 年

- **Project location:** Beijing, China
- **Project scale:** 0.1063 square kilometers
- **Design period:** 2011-2015
- **Construction period:** 2013-2017

景观设计着重处理建筑与山地环境之间的关系，将现代的生态修复技术与延续千年的中国山水美学相结合，使大型建筑群融于山水风景之中。在这一实践中，景观与建筑和谐一体，共同促成“环境友好”目标的实现。

The landscape design emphasizes the relationship between the building and the mountainous environment. It combines contemporary ecological remediation technologies with aesthetics of Chinese landscapes that have lasted for thousands of years, integrating large-scale architectural complexes into the natural landscape. In this practice, the landscape and buildings are harmonized, and together they contribute to the realization of the goal of 'environmental friendliness'.

项目区位 Project location

北京中信金陵酒店建筑群坐落于北京市平谷区西峪水库东南半山之上，建筑面积约5万平方米。酒店场地在建造前为荒废的山地果园，在果园的建造过程中对山体进行了大规模的开挖，造成原有生境的破坏。我们通过消解建筑体量，将建筑的功能性载体景观化；通过生态设计，修复被破坏的原生环境，让建筑群融入山水风景。我们致力于营造酒店所需的室外功能场地与场所意境，使之可居、可游，并产生持久的吸引力和竞争力。此项目是将当代自然环境的生态修复技术与延续千年的中国山水文化美学“仁者乐山，智者乐水”之传统相结合的实践。

生态修复技术与中国山水文化美学结合

景观设计着力处理建筑与山地环境之间的关系。通过景观设计和生态修复手法消解建筑体量，将建筑各层室外疏散楼梯、高大的挡土墙、屋顶平台和下沉院落转化为错落于山坡间的坡道、游径和观景休憩平台，与场地原生植被以及新栽植的乡土植物有机结合，并依照山、水、植被、建筑及周边环境的视线关系形成不同视觉场景的景点与观景点，与步行路径串联成系统的游赏路线和独特场景，实现“环境如画，人在画中”的意境。

酒店坐落的山坡下面原为一片淤塞的洼地、泥塘，周围生长着野生的柳树、槐树。我们梳理淤泥和植被，将水库水引至山脚，扩大了原有洼地形成景观湖体，运用传统园林艺术“借景”的手法，将西峪水库的湖景和远处山景融入场地之中。此外，我们保留了原有树木，沿湖岸种植了大量水生、湿生植物，设计了景观平台、木栈道等亲水设施，丰富了游赏路径。木质景观平台艺术化的折线造型与酒店外观呼应，木材质地则与环境更为亲和。梳理后的周边地形、保留的现有植被和生态驳岸，将景观湖与周边环境融合成一体，形成一幅天然的图画。为便于居住者的游赏活动，主体建筑北入口两侧设置了台阶，可至各层屋顶花园。散落在建筑夹缝间的19个室外庭院形成不断变化的空间片段，是人们独处、小聚、攀谈的场所。延续山地设计手法，我们让自然渗透入建筑之中，建筑与山林穿插交融。木地板沿建筑的轮廓线从地面而起，形成折面，组成挡墙、花池、树池及座椅，引导游人远离水边的危险地带，聚拢在安全且视觉良好的区域，营造现代园林让人轻松自在地游赏环境。此外，多样化的场地可以举办各种形式和规模的活动，为酒店带来经济收益，将场所转化为生产力。

弥合、重建人工环境与自然环境间的关系

在此项目中，由于山地地形的特殊性，普通雨水管不能够完全胜任山地环境的雨洪问题，而景观湖具有重要的集蓄雨水的功能。山体汇流的雨水经由建筑屋顶花园、庭院、透水路面、生态挡墙、水生植物再汇入湖中，延长了地表径流的时间，减缓了径流速度，提高了雨水的下渗率，具有一定的雨洪调节功能。湖岸的湿生植物群落可以改善水质，有利于生物的栖息和繁衍。景观湖与水库之间设计的景观坝，能有效调控湖水水位。此外，我们在湖中设置了水循环和水净化系统，净化后的水体还可以作为绿地灌溉水源。上述与绿色基础设施有关的考量，已超越了传统景观设计的范围，体现的是生态保护与“可持续”的时代要求。

由于已有设施对部分山体损伤较大。我们将生态工程与造景结合，采用生态手段对原有山体进行修复，包括运用石笼生态墙、生态护坡草毯、透水铺装、雨水花园等措施，达到固土、减少地表径流、管理雨水、过滤砂石枯叶、防止水土流失等目的。我们将当地山石碎料装填石笼，构筑生态挡墙。在挡土墙、建筑外立面种植地锦等攀缘植物，石笼挡墙内添加乡土攀缘植物及草籽组合，尽可能通过植物生长隐藏人工痕迹，让建筑掩映于自然之中。

景观环境的塑造，既具有空间维度，又具有时间维度，一年四季，循环更替。我们对树种精心选择，使园中之景可"应时而借"，时令不同，园内湖光山色也呈现出不同的景象和韵味，创造出丰富的美感和深邃的意境。

此外，我们尽可能保护原生乔灌木树林，就地造景。栽植周边原生山野植物及岩生植物在场地内进行繁衍，达到充分融入山水环境的效果。在项目实施过程中，我们为其他地方移除的植物提供庇护地，将其移植在园区适宜的位置，担当起景观设计师保护生态的责任。

由于项目涉及从山地到水边多样化的生境，为了还原并构建良好而丰富的生态系统，我们的工作不仅局限于对现场的设计，还对场地内湿地、中风化岩和微风化岩的水土保持及生态修复、地被野花演替及优化、山体排水和雨水利用等多方面进行了长期的跟踪研究，同时对现场地形的塑造、硬景的形式与位置、苗木与种植点的选择也做了反复推敲，最终确保了整个项目在景观塑造和生态修复之间实现互益共赢。

本项目弥合与重建人工环境与自然环境之间的关系，营造建筑与自然生境融为一体、相得益彰的中国山水画境，最终形成“大象无形”的审美意境。

该项目于2015年建成并投入使用，已经成为京郊以自然山水著称的度假酒店。稀缺的景观资源吸引了大量游客，在实现生态效益的同时获得了可观的经济收益。

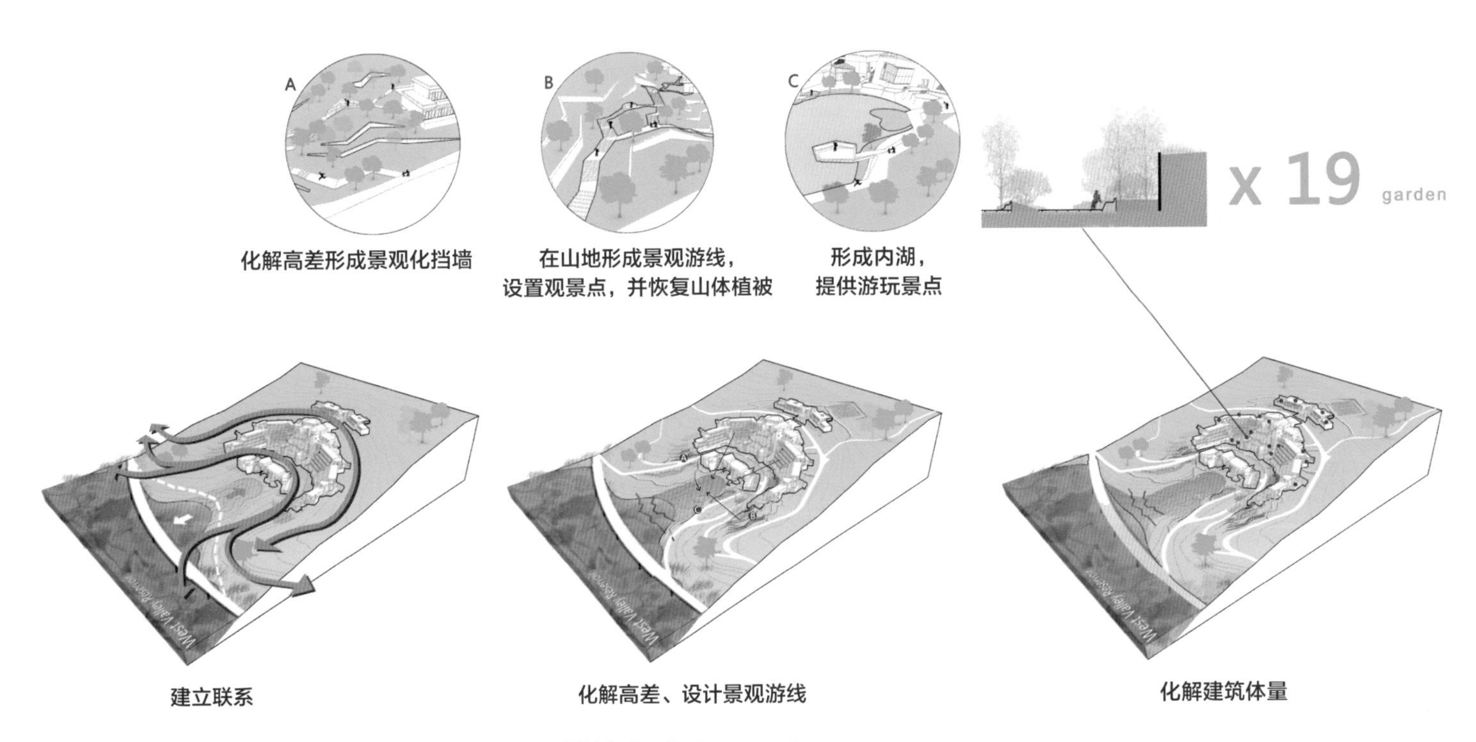

设计生成　Design generation

山地水环境示意图　Mountain water environment schematic

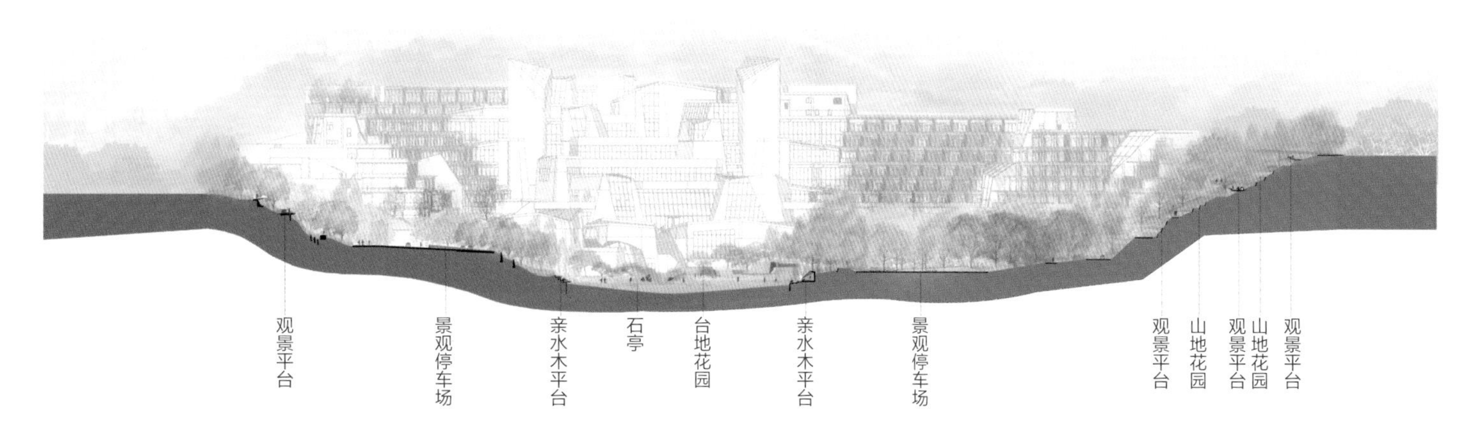

景观湖剖面图　Landscape lake section drawing

景观湖实施前后对比

Before and after comparison of the landscape lake implementation

山地环境实施前后对比　Before and after comparison of mountain environment implementation

水库水引入至山脚形成的景观湖体将西峪水库的湖景和远处山景融入场地之中
The landscape lake formed by introducing water from the reservoir into the foot of the mountain incorporates the lake and distant mountain views of Xiyu Reservoir into the site

时令不同，园内湖光山色也呈现出不同的景象和韵味

Different seasons present different views and charm of the lake and mountains in the park

景观湖边观景的休息平台

Lakeside viewing platform for landscape observation and relaxation

山地坡道　Mountain slope road

生态挡土墙将建筑掩映于自然之中　Eco-retaining walls conceal the buildings within the natural surroundings

石笼挡墙内添加乡土攀缘植物及草籽组合，通过植物生长隐藏人工痕迹
Native climbing plants and grass seed combinations are added to the gabion retaining walls, concealing artificial traces through plant growth

建筑室外庭院让自然流淌到建筑中　Outdoor courtyards of the buildings allow nature to flow into the building

营造建筑与自然生境融为一体、相得益彰的中国山水画境，形成“大象无形”的审美意境

Creating a harmonious and complementary Chinese landscape painting where architecture and natural habitat blend together, forming an aesthetic concept of " Great Form Is Beyond Shape"

生态修复与景观塑造相结合的酒店环境
Hotel environment combining ecological restoration with landscape sculpting

03 唐山凤凰山公园改造

Tangshan Phoenix Mountain Park Renovation

- **项目地点：** 中国 唐山
- **项目规模：** 0.37 平方千米
- **设计时间：** 2005-2006 年
- **施工时间：** 2006-2008 年

- **Project location:** Tangshan, China
- **Project scale:** 0.37 square kilometers
- **Design period:** 2005-2006
- **Construction period:** 2006-2008

景观设计在修复公园生态环境、提高生态服务水平的基础上，打开公园边界，建立区域慢行网络，以较低的投入为市民创造便于交往的空间，激发城市中心区活力，为老工业城市的转型发展提供契机。

The landscape design focuses on restoring the ecological environment of the park and improving the level of ecological services. It opens up the boundaries of the park and establishes a regional slow-moving network. With relatively low investment, it creates spaces that facilitate social interactions for the citizens, stimulates vitality in the city center, and provides opportunities for the transformation and development of the old industrial city.

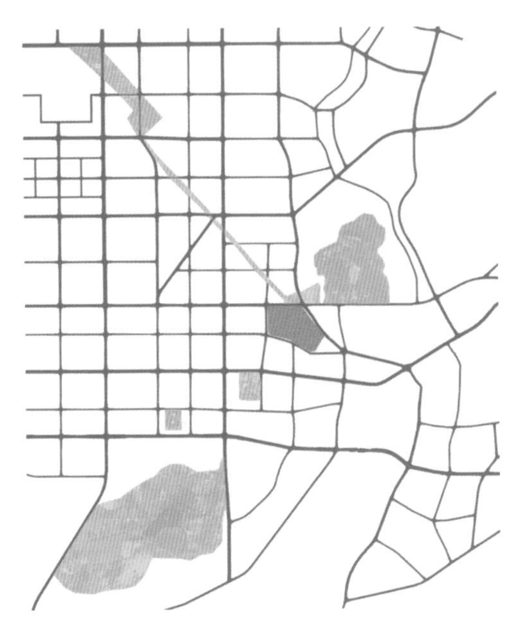

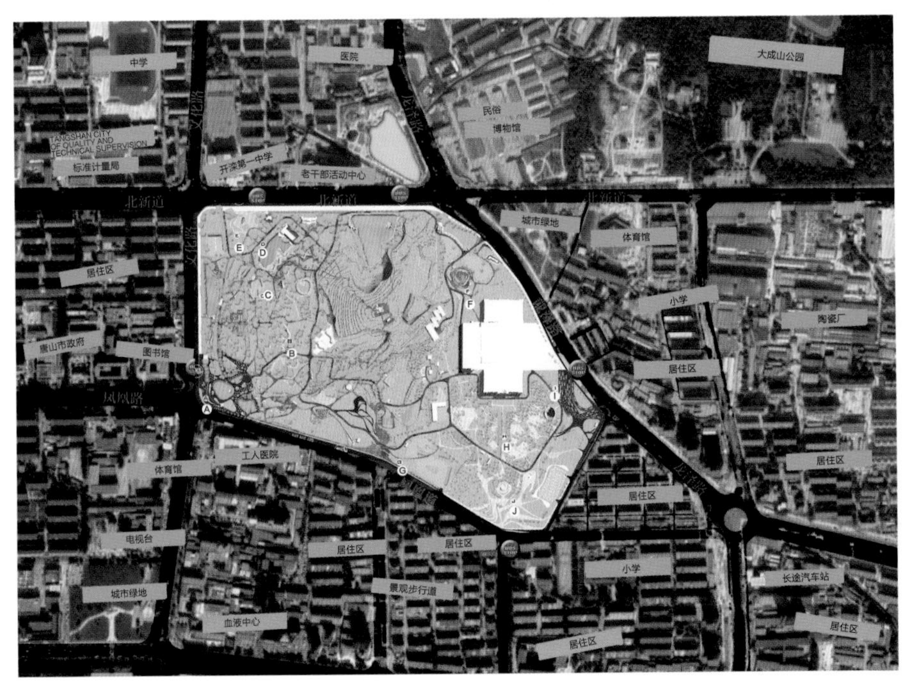

项目区位和总平面图

Project location and master plan

A南入口广场 B绿野仙踪 C听雨廊桥 D莲鱼湖 E北入口广场 F夜花园 G杏林广场 H林下空间 I银杏广场 J超级票友会

唐山市凤凰山公园位于唐山市中心，占地约0.37平方千米，始建于1956年，是唐山市建园最早的公园。凤凰山公园是唐山市民的重要社会活动场所，但由于时代变迁，老公园正在逐渐地失去活力。

修复公园生态环境

唐山凤凰山公园拥有良好的山水格局，但山体局部有滑坡和坠石的危险，水体彼此不连通，易变质和富氧化。公园内20年前修建的亭台楼阁符合人们的普遍审美，但因年久失修，早已破损，部分甚至有倒塌的危险。园内道路场地因多年无序发展显得杂乱无章，不同时期的铺装材料随意衔接，场地坑洼不平并严重不足，不少老人在裸露的土地上健身。公园拥有大片林木，但地被缺失严重，黄土裸露。

我们与专业公司合作，运用生态设计手段对裸露的山体进行防护和复绿工程，将原来存在危险隐患的区域改造成安全舒适的活动场地；联通园内水体，通过循环、曝气设备和水生植物对水质进行处理和净化，设置亲水活动空间和场地；保留园内生长良好的现状植被，在地表裸露区域适当增加活动场地，种植易养护的灌木和地被；梳理现状道路系统，选择透气透水材料和生态做法进行道路铺装，尽可能减轻环境负担；修复破损构筑物，杜绝安全隐患，增加由经济环保材料建造的休息设施，拓展游人停留空间。

现状改造，利用资源，保留场地文化基因

唐山凤凰山公园是工业化时代遗留的产物，承载了大量的历史文化信息，但与当下的城市生活疏离。在城市文化方面，我们对人文基因进行梳理，保留了有价值的活动场所和雕塑，并植入新的DNA对其进行改造，另外根据已有活动的需要增加场地的舒适度，使人们更加愿意驻足观赏和停留。重要的历史片段得到保留并与新的场景相互叠加，园内一些富有现代气息的设计则可为人们带来新鲜感，引发艺术、文化等新活动的发生。公园改造在保护市民原有生活与交流方式的同时，与新兴的生活方式融合、互动，激发新的活力。

统筹周边城市资源

公园的改造设计旨在将新的凤凰山公园及坐落在公园内的唐山市博物馆、分布在公园周边的民俗博物馆、大成山公园、体育馆、学校、干部活动中心、居住区、景观大道、图书馆、医院等城市资源和社会生活结合起来，成为城市的有机体，使公园“消解”于城市之中，成为市民的“城市客厅”，激发城市活力。

公园以“穿行”作为设计关键词，将公园边界向城市打开，穿越公园的路径将风景编织进市民的生活，公园与城市、与社会生活紧密联系。“穿行”丰富和扩展了公园的功能，使公园不再仅是传统意义的“园”，而是融于城市中的美好生活的体验场所。我们提倡市民步行或骑自行车穿越公园到达城市的各个角落，同时使其拥有更多的适宜人们交流的令人愉悦的公共空间，使公园生活成为市民日常生活的一部分。

改造后，舒缓的山坡上满是花朵

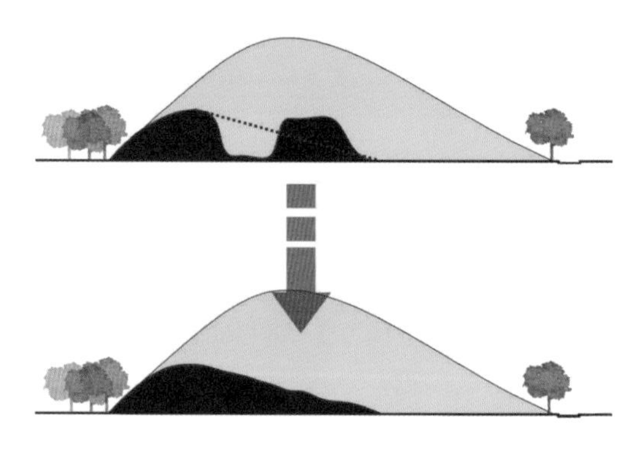

凤凰山断崖下现状地形变化过大，并有大量的建筑垃圾堆砌，环境恶劣，既不适合植物的生长，也不利于人们在此进行各种活动，加上山体裸岩风化较严重，存在安全隐患

通过对场地内地形竖向处理使地势变得较为平缓，适合人们活动，并运用加设挡土墙景观护坡等措施，解决存在的安全隐患

凤凰山断崖下空地改造前的现状照片

改造后

西南门广场改造前后对比　Before and after comparison of the southwest gate square transformation

凤凰舞会廊架改造前后对比

Before and after comparison of the Phoenix Dance Corridor transformation

废弃小火车轨道与隧道改造前后对比

Before and after comparison of the abandoned small railway track and tunnel transformation

改造前

改造后

改造后

林下空间改造前后对比

Before and after comparison of the understory space transformation

公园内丰富的市民活动

Abundance of civic activities in the park

郊野公园
体系建设

CONSTRUCTION OF SUBURBAN PARK SYSTEMS

郊野公园起源于英国，是指位于城市近郊、有着良好自然景观、郊野植被及田园风貌，并以休闲娱乐、生态保护为目的的郊外旅游休憩区域。郊野公园作为衔接城市与乡村生态系统的中间环节，对统筹城乡发展、改善生态环境具有不可替代的作用。

Suburban parks, originating from the UK, refer to recreational areas located in the outskirts of cities with good natural landscapes, rural vegetation, and pastoral scenery. They are intended for leisure, entertainment, and ecological conservation purposes. As an intermediate link between urban and rural ecosystems, suburban parks play an irreplaceable role in coordinating urban-rural development and improving the ecological environment.

中国福建漳州地处全球黄金气候带，四季常绿，花果累累，农耕历史悠久，田园风光优美，具有建设郊野公园的天然优势。九龙江是漳州的母亲河，城市依江而建，农田沿岸而作，千百年来江、城、人、景和谐发展。漳州老城区位于九龙江北岸，新的城市总体规划将城市格局拓展为“一江两岸”，这种城市化或将改变九龙江两岸现有的自然景观风貌。为避免城市建设发展对环境的破坏，同时满足市民日益增长的户外休闲需求，漳州市积极探索城市与自然互动、城市与乡村统筹的生态文明建设途径，聘请著名规划师黄晶涛先生带领天津愿景公司团队完成了具有前瞻性的漳州郊野公园体系规划工作。

在此规划基础上，无界景观陆续完成了九龙江畔西溪湿地郊野公园、漳州滨江生态公园（天宝段）、漳州南山生态文化园、漳州云霄县南湖湿地生态园，以及漳州平和县九龙江支流（高际溪）、漳州华安真武山的景观设计工作，实践区域从漳州中心城区逐渐拓展至下属县（市），郊野公园与风景河道、溪流湿地、自然山体、田园果林、民居古迹等现状资源有机结合，加强了城乡统筹，优化了发展格局，在展现田园秀色的同时促进本地产业转型，形成一幅百姓安居乐业的生态宜居画卷。

Zhangzhou, located in the global golden climate zone, boasts evergreen seasons, abundant flowers and fruits, a long history of agriculture, and beautiful pastoral scenery, making it naturally suitable for the construction of suburban parks. The Jiulong River is Zhangzhou's mother river, around which the city has developed, with farmland along its banks. For thousands of years, the river, the city, the people, and the scenery have harmoniously developed together. The old town of Zhangzhou is situated on the northern bank of the Jiulong River. The new urban master plan expands the city's layout to "one river, two banks," which may alter the existing natural landscape along both sides of the river. To avoid environmental damage caused by urban development while meeting the increasing outdoor leisure needs of the citizens, Zhangzhou actively explores the path of ecological civilization construction that involves interaction between the city and nature and the coordination of urban and rural development. Led by the renowned planner Mr. Huang Jingtao, the team from Tianjin Vision Company has completed the forward-looking planning of the Zhangzhou suburban park system.

Based on this plan, View Unlimited Landscape Architects Studio has completed landscape design work for "Jiulong River Xixi Wetland Suburban Park," "Zhangzhou Binjiang Ecological Park (Tianbao Section)," "Zhangzhou Nanshan Ecological Cultural Park," "Zhangzhou Pinghe County Jiulong River Tributary (Gaoji Creek)," "Zhangzhou Yunxiao County Nanhu Lake Wetland Ecological Park," and "Zhangzhou Hua'an Zhenwu Mountain." The practice area has gradually expanded from the central urban area of Zhangzhou to its subordinate counties and cities. The suburban parks are organically integrated with existing resources such as scenic rivers, streams, wetlands, natural mountains, pastoral orchards, and historical dwellings. This strengthens the coordination between urban and rural areas, optimizes the development pattern, showcases the beauty of the countryside and promotes local industrial transformation, creating an ecological and livable picture for the well-being and livelihood of the people.

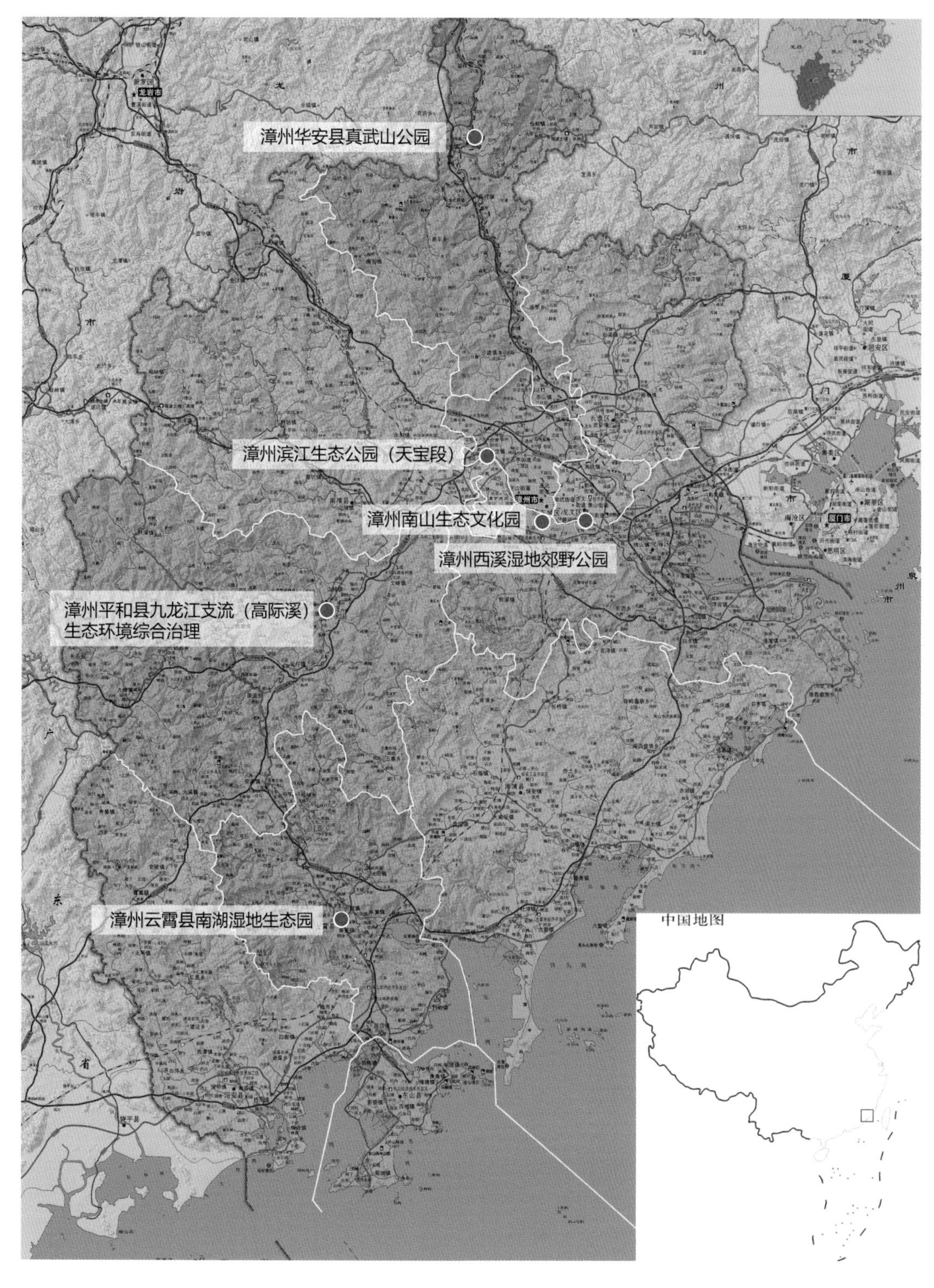

无界景观陆续完成了九龙江畔西溪湿地郊野公园、漳州滨江生态公园（天宝段）、漳州南山生态文化园、漳州平和县九龙江支流（高际溪）、漳州云霄县南湖湿地生态园、漳州华安真武山的景观设计工作，实践区域从漳州中心城区逐渐拓展至下属县（市）

View Unlimited Landscape Architects Studio has completed landscape design work for "Jiulong River Xixi Wetland Suburban Park," "Zhangzhou Binjiang Ecological Park (Tianbao Section)," "Zhangzhou Nanshan Ecological Cultural Park," "Zhangzhou Pinghe County Jiulong River Tributary (Gaoji Creek)," "Zhangzhou Yunxiao County Nanhu Wetland Ecological Park," and "Zhangzhou Hua'an Zhenwu Mountain." The practice area has gradually expanded from the central urban area of Zhangzhou to its subordinate counties and cities

01 漳州郊野公园体系建设与漳州西溪湿地郊野公园景观设计

Construction of the Zhangzhou Suburban Park System and Landscape Design of Zhangzhou Xixi Wetland Suburban Park

- **项目地点：** 中国 漳州
- **项目规模：** 设计面积 1.4 平方千米；规划面积约 100 平方千米
- **设计时间：** 2011-2012 年　　**施工时间：** 2012 年
- **郊野公园体系规划：** 天津愿景城市开发与设计策划有限公司
- **景观设计合作单位：** 天津愿景城市开发与设计策划有限公司

- **Project location:** Zhangzhou, China
- **Project scale:** Design area:1.4 square kilometers; Planning area about 100 square kilometers
- **Design period:** 2011-2012　　**Construction period:** 2012
- **Suburban park systems Planning:** Tianjin Vision Urban Development and Design Planning Co., Ltd
- **Landscape design cooperation unit:** Tianjin Vision Urban Development and Design Planning Co., Ltd

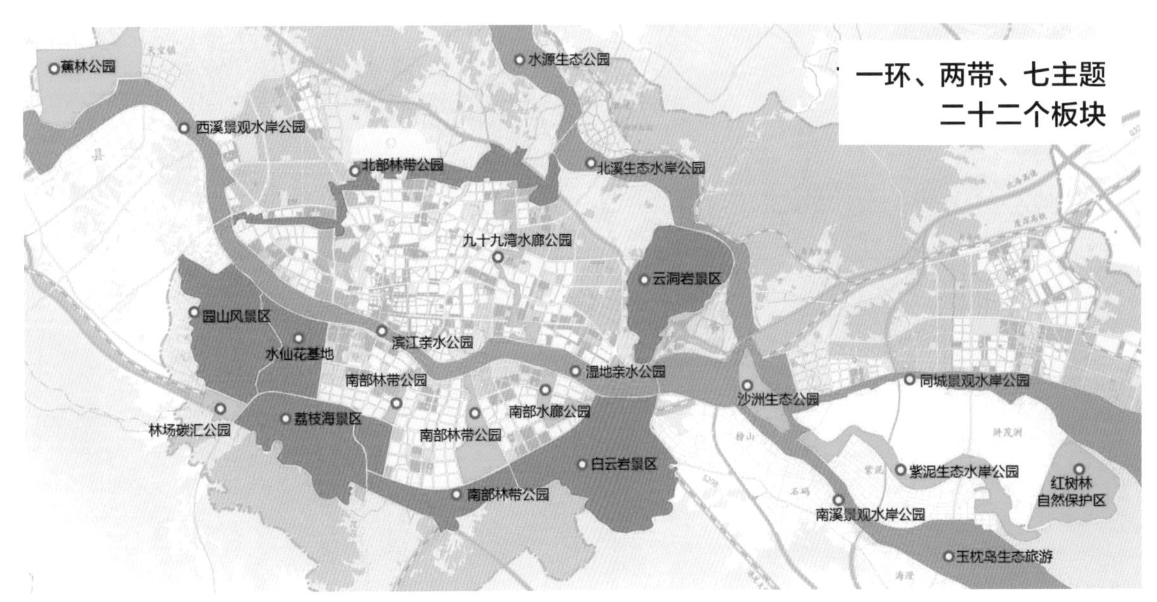

将城郊的绿色山林、田园、湿地水体、自然保护区、风景区等纳入郊野公园体系，形成具有漳州特色的“翡翠项链”，构建“城乡一体复合生态系统”

Incorporating suburban green forests, countryside, wetlands, nature reserves, and scenic areas into the suburban park system, forming a "jade necklace" with Zhangzhou characteristics, and constructing an integrated urban-rural ecological system

郊野公园的景观设计最大限度地保护与利用原有地形、水系、植被、村落、道路等资源，展现漳州优渥的自然山水环境，串联城乡绿色空间，拉近市民与自然之间的距离。

The landscape design maximizes the protection and utilization of existing resources such as topography, water systems, vegetation, villages, and roads. It showcases Zhangzhou's abundant natural landscapes, connects urban and rural green spaces, and bridges the gap between citizens and nature.

“园在城中，城在园中，城园一体，园城共荣”的郊野公园体系

项目地处漳州市九龙江北岸，横跨漳州中心城区，现状沿江绿地的生态环境遭到一定破坏。我们践行“全城郊野公园行动规划”，整合物质与非物质资源，带动城市生态系统、绿地系统、人文系统以及旅游系统的全面提升。以生态学原理为指导，打通“廊道”，梳理串联散布的现状“斑块”，将城郊的绿色山林、田园、湿地水体、自然保护区、风景区等纳入郊野公园体系，保持和维系区域生态平衡，形成具有漳州特色的“翡翠项链”，构建“城乡一体复合生态系统”。

我们借鉴国内外郊野公园建设的先进经验，推行“以水为脉、以绿为韵、以文为魂”三大策略，突出漳州山、水、林、田自然景观以及特色文化，规划形成覆盖漳州中心城区约100平方千米的郊野公园体系，建设“田园都市、生态之城”。漳州郊野公园体系包含22个特色板块，形成“一环、两带、七主题”的总体结构。连通一环，打通城区南北林带，串联圆山、荔枝海等景区，构成围绕中心城区的生态绿环，既是提供生态涵养的绿色空间，又是隔离工业的生态屏障，并界定城市发展边界。突出两带，依托九龙江西溪、北溪两条水脉，结合九十九湾等水廊，保留河流原有生态廊道，打造城市生态水网，提供自然亲水空间。确定七大主题，根据现状不同类型资源划分为水岸公园、水廊公园、山林公园、林带公园、农林公园、岛屿公园、湿地保护区公园，并制定不同类型的景观营造策略，保护并展现漳州自然生态原有的景观多样性，丰富漳州郊野公园系统的景观体验。

我们追求郊野公园与城市整体发展的实时互动，发挥在生态涵养、环境调节、形象展示、休闲游憩、文化传承等方面的多重功能，最终形成“园在城中，城在园中，城园一体，园城共荣”的独特风貌，统筹城乡发展，保障生态、社会、经济效益的协调与兼容，实现人与自然的和谐共生。

注重城市水脉保护与提升的西溪湿地郊野公园

我们将九龙江水脉定位为城市生态、文化、休闲、生活综合性廊道，摒弃大改大建、人工痕迹过多的河道改造方式，打造大绿野趣的沿河空间。设计紧密结合现状条件，并在实施过程中不断进行优化调整，实现了对于原有地形、水系、植被、村落、道路等最大限度的保护与合理利用。

公园提供步行、慢跑、自行车骑行、游船等多种游览形式，设计多样化的休闲、运动场所，鼓励全民健身，拉近市民与自然之间的距离。设计自然观察径，串联不同植物和鸟类的观察点，打造自然教育基地。结合漳州传统民俗，为九龙江龙舟赛设置龙舟下水点及观赛场地，传承龙舟文化。保留沿岸现状村落，推进产业转型，由农业向旅游服务业转变。结合村落中的庙宇、戏台、碑刻等进行公共空间的提升改造，为民俗文化活动提供场所，营造村民幸福宜居，游客在地体验的“都市桃源”。

建成的九龙江郊野公园已成为全民健身、传统民俗、科普宣教的优良场所，影响并改变着市民的生活方式，推动旧有城市记忆与新兴生活场景的交融、共生。

保留现状石桥，展现九龙江两岸山水风貌
Preserving the existing stone bridges to showcase the landscape of both sides of Jiulong River

梳理现状滩地水系，丰富空间变化
Organizing the existing beach water system to enrich spatial variations

从九龙江郊野公园远望龙文塔
Viewing Longwen Tower from Jiulong River Suburban Park

湿地岛上植物繁茂，展示了湿地生态系统　Lush vegetation on the wetland island, showcasing the wetland ecosystem

保留江畔原生态景观　Preserving the original ecological landscape along the riverbank

开辟滞洪区湿地，提升景观风貌
Developing flood detention wetlands to enhance the landscape aesthetics

现状荔枝园内的土路改造为林荫园路
Transforming the existing dirt road in the lychee orchard into a tree-lined path

补植地被花卉，营造花堤景观
Planting flowers on the embankment to create a flower dike landscape

林下开放空间　Open understory space

市民健身空间　Fitness space for citizens

02 漳州滨江生态公园（天宝段）景观设计

Landscape Design of Zhangzhou Binjiang Ecological Park (Tianbao Section)

- **项目地点：**中国 漳州
- **项目规模：**0.563 平方千米
- **设计时间：**2016-2017 年
- **施工时间：**2018 年

- **Project location:** Zhangzhou City, China
- **Project scale:** 0.563 square kilometers
- **Design period:** 2016-2017
- **Construction period:** 2018

滨江生态公园延续郊野公园“大绿野趣”的定位，保护九龙江西溪自然景观资源，持续提升九龙江环境质量，保护区域生态环境健康发展。

The Binjiang Ecological Park continues the positioning of suburban parks with a focus on natural beauty and enjoyment of the countryside. It protects the natural landscape resources of the Jiulong River's West River and continuously improves the environmental quality of the Jiulong River, ensuring the healthy development of the regional ecological environment.

漳州滨江生态公园天宝段位于九龙江西溪北岸、迎宾西路（G319）南侧，距漳州市中心约10千米，占地面积0.563平方千米，是漳州城市郊野公园体系“两带”之一的“九龙江西溪带”最西端的水岸公园。

公园地块沿九龙江北岸和迎宾西路平行延展，长度约为4.7千米。作为漳州城市郊野公园的组成部分，滨江公园将延续郊野公园大绿野趣的定位，保护九龙江西溪自然景观资源，梳理、串联散布的现状“斑块”，纳入山、水、林、田的自然景观及特色文化，打造原生态的滨江“绿色廊道”，并以公园建设为契机，提升九龙江环境质量，保护区域生态环境健康发展。

风景融入日常生活，构建优美和谐的山水城市

根据天宝镇镇域规划，滨江公园西部紧邻镇区核心，北部则是镇域的综合服务区，周边用地以居住和商业为主，也包含了一些行政办公、教育用地等，地块内有多条区域交通干道交错。相对于一般服务周边的城市公园，带状公园与城市的交界面更长，关系更为紧密。沿西溪展开的狭长公园服务于天宝镇，是城市发展轴、滨水绿化带、绿化通廊的交汇节点，是连接天宝与漳州中心城区的绿色廊道。设计追求公园与城市整体发展的实时互动，并衔接周边资源，巧妙借景，实现“园在城中，城在园中”，让优美风景融入市民的日常生活。

滨江公园所沿江段是九龙江西溪最美丽的转弯，有着茂密的竹林景观。景观设计尊重原有地形地貌，保护西溪自然的原生态驳岸，梳理现状水系，修复已被破坏的竹林，形成连续的滨江竹林景观。“宝峰飞翠”和“圆峤来青”是漳州市“古八景”中的两处，宝峰指公园北部的天宝大山，该山属于福建第二大山脉，山上保留大量原生态的森林，“圆峤”指九龙江南岸的圆山，其东南山麓是驰名海内外的水仙花产地。设计运用借景的手法，构建山、水、林、田的优美视觉通廊，并注重功能互补，助力区域的协调发展。

一体化营建文化基础设施，塑造功能复合的弹性公共空间

漳州地处闽南，具有浓厚的地域民俗文化，其中饮食文化浓厚，重养生、饮茶，天宝是世界级文学大师林语堂先生的祖籍地。场地东部珠里村建有占地0.447平方千米的林语堂文化园；场地东北则是茶铺村畲族特色村寨。设计突出地域文化特色，链接天宝镇特色人文资源，通过设计形成空间丰富、景观优美的开放空间，为当代文创活动搭建发展平台，为众多“创客”提供创作、交流、展示、宣传的场所，激发天宝镇创意产业活力，带动区域经济发展。

项目结合漳州的养生文化，融入畲族独特的体育活动，发扬与民共建的和谐精神，利用滨江竹林优美的环境，串联场地内运动场地，形成内涵丰富的健身公园，方便市民开展竹林太极、瑜伽、畲族武术、秋千等健身活动。户外健身空间通过慢行线路融入省级绿道网络，与西溪郊野公园整体打造全国最具特色的滨江户外健身基地。

建成后的漳州滨江生态公园不仅为周边居民提供了一个休闲健身的场所，也为天宝镇打开了一扇绿色活力的门户，更为漳州建设“田园都市、生态之城”增添了独具特色的一笔。

改造前

改造后

建成前后对比照片　Before and after comparison photos of the construction

利用现状堤，面向九龙江最美转弯设计景观平台及码头，游客在等船时也能欣赏美景

Utilizing the existing embankment, a scenic platform and dock facing the Jiulong River is designed to allow tourists to enjoy the beautiful scenery while waiting for boats

梳理联通现有低洼地，结合湿生植物种植，营造大绿野趣的游赏氛围。穿行其中的木栈道是欣赏江景和对岸竹林的最佳游赏路线

By rearranging the existing low-lying areas and combining them with wetland plantations, we aim to create a large green and leisurely atmosphere for appreciation. The wooden walkway running through it is the best route for enjoying the river view and the bamboo forest on the opposite bank

改造前

建成后的实景照片
Real-life photos of the completed project

建成后的实景照片

Real-life photos of the completed project

利用竖向高差和竹林设计沿江观景木栈道，为游人提供欣赏江景更好的视角

By utilizing the vertical height difference and bamboo forest, a riverside wooden walkway is designed to provide visitors with a better view of the river

以最大限度地保护现状竹林为前提，利用现状路在林中开辟新园路，组织合理的游览路线，为游人提供舒适的竹林漫步空间

Taking maximum consideration of protecting the existing bamboo forest, a new park road is opened within the forest using the existing roads. This provides visitors with a comfortable space for strolling amidst the bamboo forest

建成后的实景照片
Real-life photos of the completed project

依托茶铺村的畲族文化资源，利用现状开阔地设置畲族文化花园，布置戏台、摄影工作室、小卖部、茶室等服务设施，既满足市民休闲需求，又可以承担节庆活动
Leveraging the cultural resources of the She ethnic group in Chapu Village, an ethnic cultural garden is created in the open area. This garden includes a stage, photography studio, small shops, and teahouses, which will not only meet the leisure needs of the citizens but also host festive activities

利用石笼墙稳固现状裸露的土坎，形成台地空间，上层空间作为集散场地，下层滨水区域建成观景码头
Stabilizing the existing bare slopes with stone gabions and creating a platform area, the upper space serves as a gathering place, while a viewing dock is built in the lower riverside area

公园内的休闲服务设施　Recreational facilities within the park

03 漳州南山生态文化园景观设计

Landscape Design of Zhangzhou Nanshan Ecological Cultural Park

◎ **项目地点：** 中国 漳州

◎ **项目规模：** 0.53 平方千米

◎ **设计时间：** 2017 年

◎ **施工时间：** 2017 年

◎ **Project location:** Zhangzhou City, China

◎ **Project scale:** 0.53 square kilometers

◎ **Design period:** 2017

◎ **Construction period:** 2017

景观设计以“千年文化，一脉相承”为目标，通过漳州千年历史文化轴线串联山、江、湖、田、寺、城，让山水风光融入城市，营造“碧水环青山，花海拥古刹，登高望古城，乐活享南山”的自然诗意的人文风光，形成集湖光山色、休闲健身、文化交流于一体的城市新空间。

The design aims to embody the concept of "millennial culture, continuous heritage" by connecting mountains, rivers, lakes, fields, temples, and the city along the millennium historical and cultural axis of Zhangzhou. The goal is to integrate the natural beauty into the urban fabric, creating a poetic and humanistic landscape where "crystal-clear waters surround green mountains, flower seas embrace ancient temples, ascending heights offer views of the ancient city, and people enjoy their joyful and serene lifestyle offered by Nanshan Mountain." This design will form a new urban space that combines the beauty of lakes and mountains, leisure and fitness activities, and cultural exchanges.

项目区位图

Project location map

南山生态文化园位于九龙江的南侧，与漳州老城隔九龙江相望。相对于九龙江北的唐宋古城历史文化保护区，这里并不是历史积淀最集中的区域，但这里坐落着丹霞山、南山，以及始建于唐朝的南山寺，山体和寺庙——自然环境和人文景观的组合，构成了漳州历史轴线的终点，也是漳州历史上的南大门。

重塑漳州山水城景

南山生态文化园设计范围红线内总用地面积0.53平方千米，包括现状水域面积0.067平方千米。南山和丹霞山为两座连峰的小山，均不高于50米，山谷之间为现状村落，村落的生长逐渐蚕食了大量绿地。本项目会将村落迁出，人工建设的痕迹也将得到修复，然后通过建设公园绿地，梳理、净化水系，重塑山水相融的景色。

在两山景色的重塑和提升上，设计团队循序渐进地进行了场地现状植被的梳理，场所文化内涵的发掘。南山和丹霞山现状植被茂密，有大面积的荔枝林，古有“南山秋色（南山寺赏秋）”“朝丹暮霞（丹霞山观日出和日落）”两大名景。在南山景色的营造上，依据元朝安溪主簿林广发诗句“翘首城南土，悠然见此山。竹藏秋雨暗，松度晚风寒。佳色催黄菊，晴光上翠峦”，通过竹林、秋菊，以及秋色叶植物点景，营造素雅的“南山秋色”氛围，与山脚下的南山寺融为一体。在丹霞山景色的营造上，一方面修复植被，另一方面，将重点放在林下空间的营造上，丰富多彩的林下课堂和健身活动区让丹霞山充满活力。在制高点设计城市观景台，北望漳州古城，南探七首岩，西观圆山，东衔南山，重塑朝丹慕霞的观景地。此外，南山和丹霞山被现状村落所隔开，未来也被规划车行路所隔断，为了加强生态效益和景观体验的连贯性，通过道路连通两山，构成南山文化园的标志性山体景观。

在水景的营造上，首先梳理水系，保证场地水体的连通性；其次提升水质。场地内水体现状污染比较严重，需要截断外源设置污水处理设施，通过曝气机增加水体的流动，以及利用生态手段提升水体自净能力，最后完成水景的营造工作。一侧水体结合南山寺营造湖体景观，另一侧在现有鱼塘的基础上改造提升，形成湿地景观。既有开敞的湖面，富有禅意的莲花与南山寺相伴，又有错落的湿地、田埂，可以穿行其中，构成丰富有趣的水景体验。

打造漳州千年历史轴线

南山生态文化园有着丰富的历史沉寂，总结起来可以用“七古”概括——古轴线、古景色、古寺庙、古驿站、古桥、古井、古街巷。设计对这些历史文化资源进行分类梳理，确定哪些是需要保留的历史建筑、景观，哪些是需要提升的环境，哪些是已经消失并需要重塑的历史景观，因地制宜地进行保留、改造、提升、重建等工作，并赋予更加丰富的功能，让这些古老的文化资源可以走进市民的日常生活。

策划南山文化体验、慢运动健身、赏三角梅三大主题活动。通过合理的路径设置，将保留的古巷、古寺庙、古井和戏台串联成民俗文化体验游径，提供闽南非物质文化遗产的教育与体验场地。复兴丹霞古驿站，形成建筑组团，承载丰富的商业文化活动，构成南山文化园的商业文化中心。结合自然山水在整个园区引入慢运动、慢生活的生活理念，发展健身休闲产业。依托南山寺一侧的三角梅大观园，拓展三角梅产业链，形成集展示、餐饮、销售为一体的体验式花海。此外灯光秀设计以及丰富的夜晚活动让南山成为漳州人夜游的新地标。

建成后的南山生态文化园呈现出绿水环青山，花海拥古寺的美景。九龙江上的南山桥将漳州老城和南山生态文化园联系在一起，一面是“南山秋色”“朝丹暮霞”的新景观，一面是充满活力的漳州老城，两者无缝衔接。本项目再现了漳州的千年历史轴线，也让人们对于漳州文化、休闲生活的体验从九龙江的北岸延伸到南岸。

梳理现状水系，形成湿地景观

Reorganizing the existing water system to create a wetland landscape

利用水利设施修建观景台阶

Constructing a viewing terrace using hydraulic facilities

丰富的湿地水景体验
Rich wetland water landscape experience

沿路设置的观景台阶　Viewing steps set up along the route

公园休闲活动场地
Leisure activity areas within the park

优美的风景路
Beautiful scenic road

04 漳州平和县九龙江支流（高际溪）生态环境综合治理与环境景观设计

Ecological Environment Comprehensive Management and Environmental Landscape Design for Pinghe County Jiulong River Tributary (Gaoji Creek)

- **项目地点：** 中国 漳州
- **项目规模：** 0.546 平方千米
- **设计时间：** 2017 年
- **施工时间：** 2022-2023 年

- **Project location:** Zhangzhou City, China
- **Project scale:** 0.546 square kilometers
- **Design period:** 2017
- **Construction period:** 2022-2023

景观设计以一体化生态营建发挥高际溪在未来城市新区中的生态功能，融入健身和教育功能，创造边玩边学的户外空间，在玩耍中加强居民对家乡文化的认知，引导居民展开环境共建活动，凝聚共同回忆，获得认同感和归属感。

The design aims to utilize integrated ecological construction to enhance the ecological functions of Gaoji Creek within the future urban area. It incorporates fitness and educational elements to create an outdoor space where people can play and learn simultaneously. Through recreational activities, residents can deepen their understanding of local culture, participate in environmental co-creation activities, foster shared memories, and develop a sense of identity and belonging.

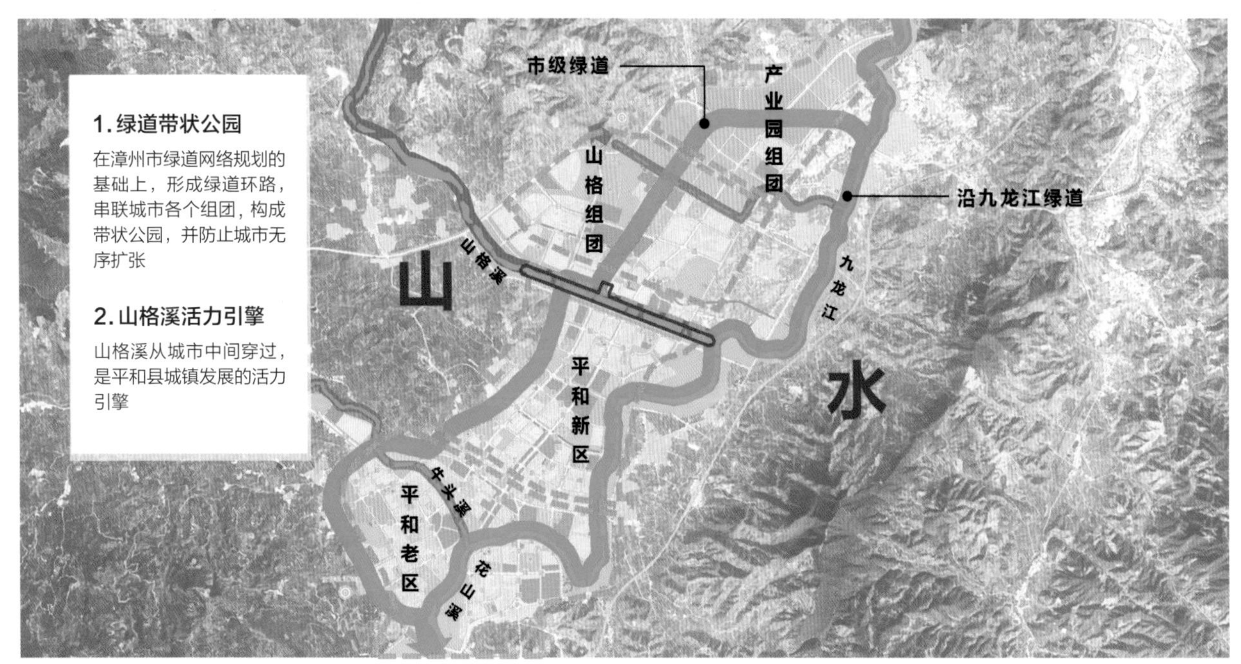

项目区位
Project location map

平和县地处福建省漳州西南部，与闽粤两省八县毗邻，素有“八县通衢”之称，是福建省重点侨乡之一，也是台胞的重要祖籍地。漳州平和县九龙江支流高际溪，位于山格组团和平和新区之间，是未来城市发展的中心，设计定位为新区活力水岸。水岸两侧大部分区域规划为居住用地，有配套的商业、学校和医疗。

一体化营建生态系统，统筹自然、人文、休闲和教育功能

在高际溪两岸的生态治理上，首先要做的是高际溪及两岸生态系统的营建，其次在场地设计内融入亲水休闲、健身养生、自然教育等系统，提升人们的幸福指数，营造美好的居住使用体验。此外，更为重要的是让居民参与到公园环境的建设中，从而收获更多的责任感与归属感。

高际溪原为排洪渠道，在城市中承担了河道排洪的作用，河道笔直，除局部硬质堤岸外，大部分为自然驳岸。一体化营建的生态系统统筹考虑水岸生态环境与水质提升、蓄滞能力提升、生物多样性营造，以及可再生资源的利用等多项生态问题。首先，保证高际溪的行洪排涝功能以及在城市中承担的海绵系统的角色得以延续。利用河道两侧的洼地作为季节性汇水区域，实现蓄存雨水、净化雨水的功能，并形成季节性景观；设置植草沟，组织地表径流，疏导净化雨水。其次，结合生态系统的改造以及营建工作，配套相关自然教育设施，普及生态保护的相关知识，增加趣味性的学习体验。

高际溪两侧以居住、学校用地为主，是承载周边居住者、使用者日常生活的家门口的滨水公园。因此景观设计更加重视场地使用功能的多样性，以满足不同年龄市民的需要，鼓励市民在场地内进行各类有益身心健康的文体活动，创造边玩边学的户外空间，让孩子们在玩耍中提升对家乡文化的认知。

（1）建立适合不同人群的市民健身系统

当地最受欢迎的体育活动是太极拳和广场舞，因此我们在河道两岸设计了不同尺度的场地，满足多种规模的广场舞和太极拳活动的要求。此外，设置专业的运动场地、慢跑径和无器械健身场地。专业健身场地以篮球场、羽毛球场、儿童运动场为主。慢跑径沿河岸布置，沿途环境优美，并形成多个不同长度的环线。无器械健身场地沿河道两侧分布，依托景观设施、户外家具设置，作为专业健身场地的补充，推广随时随地的健身方式。

（2）建立边玩边学的公众教育系统

开放的公共空间内包含自然的更替、有趣的体验，以润物细无声的方式持续发挥公众教育的功能。高际溪的公共教育系统侧重两个方向：一是结合生态设施和共建活动，建立户外自然课堂，二是通过五感体验理解林语堂先生的生活。户外自然课堂包含小气候课堂、有机柚田课堂、水生态水利课堂、山花植物课堂等，分布在河道两岸的公共空间中，受众在游玩过程中收获不同类型的自然知识。在语堂长廊中，以特色铺装方式将林语堂先生翻译的《道德经》娓娓道来。

（3）通过居民共建的乡村建设方式培养归属感，让山花开满高际溪

在高际溪沿岸景观设计中，我们发动居民收集当地适宜生长的山花山草种植在道路两侧，并结合铺装对花草名称进行标识，开满鲜花的高际溪将成为平和县的自然地标，凝聚人们的共同回忆，成为人与环境、人与人情感联系的纽带，从中获得对家乡的自豪感、认同感与归属感。

漳州平和县九龙江支流的景观设计项目是统筹建设的又一次实践，除多专业协作、一体化设计外，还鼓励居民自发营造，共同创造居住环境，成功打造出“望得见山、看得见水、记得住乡愁”，承载人们的共同记忆，在时间中不断生长变化的景观形态。

鸟瞰效果图 A bird's-eye view rendering

山花开满 高际溪

鼓励居民在水岸种植收集于本地的花籽，让山花开遍高际溪

Encouraging residents to plant and collect flower seeds along the water's edge, allowing mountain flowers to bloom throughout Gaoji Creek

一体化营建生态系统

Building an integrated ecological system

休闲广场举办市民健身活动　Hosting citizen fitness activities in the leisure plaza

柚田茶亭乐享田园生活　Tea Pavilion in pomelo orchard for enjoying a pastoral life

滨水休闲步道观赏湿地风光
Enjoying the wetland scenery along the riverside promenade

05 漳州市云霄县南湖湿地生态园景观设计

Landscape Design of Yunxiao County Nanhu Wetland Ecological Park

- **项目地点：** 中国 漳州
- **项目规模：** 0.745 平方千米
- **设计时间：** 2017 年
- **施工时间：** 2018-2019 年

- **Project location:** Zhangzhou City, China
- **Project scale:** 0.745 square kilometers
- **Design period:** 2017
- **Construction period:** 2018-2019

南湖湿地生态园是漳州市云霄县县城新区的核心绿地，下游是我国纬度最高、面积最广、植被种类最多的红树林天然群落，天然湿地条件优越。设计以“生态优先、最小干扰”为策略，保护利用天然湿地资源，构建湿地水生态系统，打造云霄南湖生态旅游品牌。

The Nanhu Wetland Ecological Park project is the central green area of the new urban district in Yunxiao County. Downstream lies the most diverse mangrove natural community in China, with the highest latitude and widest area. The natural wetland conditions are excellent. The design follows the principle of "ecological priority and minimal disturbance" to protect and utilize the natural wetland resources. It aims to create a wetland aquatic ecosystem and establish the brand of eco-tourism for Nanhu Wetland in Yunxiao.

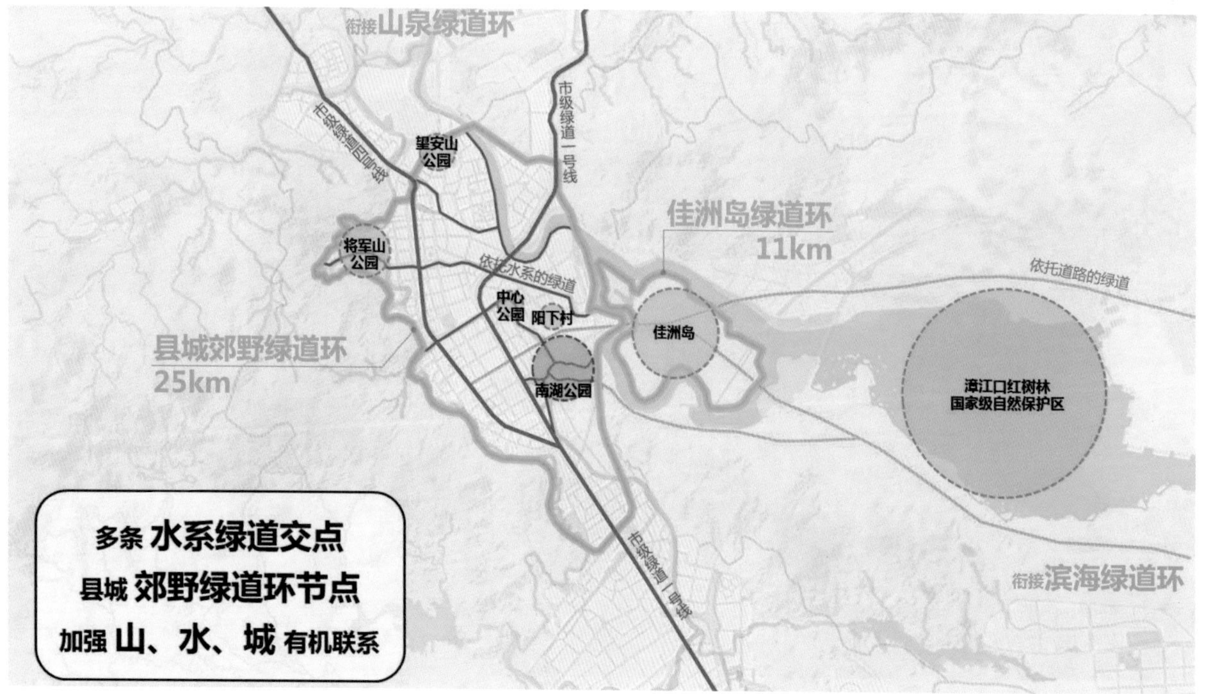

以水为脉，完善绿地网络，打造县城郊野绿道环，加强山、水、城之间的联系

Improving the green space network by focusing on water and creating a suburban greenway ring in the county that enhances connections between mountains, water, and the city

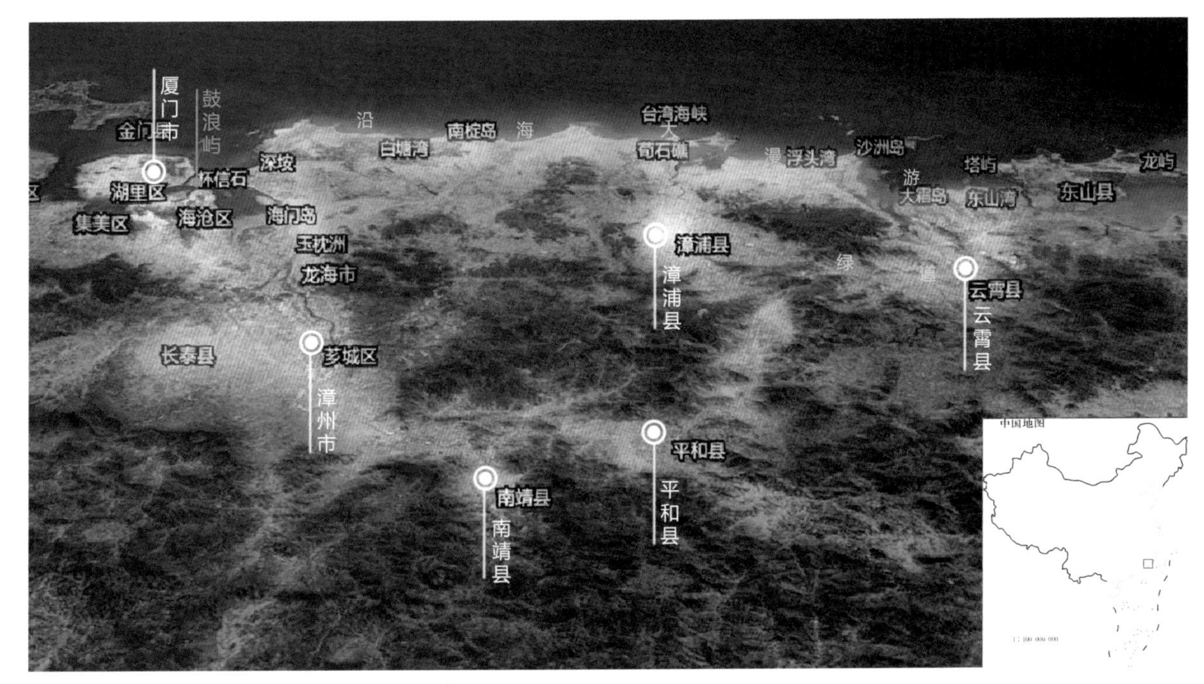

积极参与构建闽南金三角、山、江、海、城大网络，促进漳厦一体化

Actively participating in the construction of a large network connecting Jinshan Mountain, rivers, seas, and cities in southern Fujian, promoting the integration of Zhangzhou and Xiamen

南湖湿地生态园项目所在的云霄县地处福建省漳州市南部，素有“开漳圣地”之称谓，与台湾省隔海相望。云霄县倚山傍海、溪涧纵横，境内大小河流近20条，汇集于漳江，经过城区向东南入海。位于云霄县漳江入海口的“漳江口红树林国家级自然保护区”总面积23.6平方千米，是福建省唯一的国家级湿地自然保护区、国际重要湿地，是我国纬度最高、面积最广、植被种类最多的红树林天然群落。

南湖湿地生态园项目位于云霄县老城区东南部，漳江口红树林保护区的上游，南临山美山——大步山绿楔，东部与佳洲岛隔南江相望，山美、前涂、蒲南、汀仔洋四溪在项目地块内交汇，有着优越的天然湿地条件。

“生态优先、最小干扰”的湿地水生态系统设计

湿地系统与其他生态系统一样，由生物群落及与之相互作用的环境构成。湿地景观设计应综合考虑各个因素，以统筹协调、整体和谐为主旨，才能实现生态设计的目的。设计师从原有环境的调查入手，包括自然环境、社会条件、居民情况的调查，以“生态优先、最小干扰”为策略，力求在满足人居需求的同时，保持自然生态不受破坏，使人与自然融洽共存，从而保持城市湿地系统的和谐与完整。

设计范围内有多条溪流汇入山美溪，然后自西向东流入南江，汇入漳江。现状溪流大部分保持天然湿地状态，场地内有大面积的田地、荷塘，少量林地，局部有硬质场地及建筑。设计依托天然湿地，保留原有的景观元素，将公园、水景界面向城市开敞，使林地、田园界面自然衔接过渡，并提升“水景观”，保护“水生态”，守住“水安全”。

项目做到不开挖、少填方，尊重自然地形水系，提取乡土植物设计语汇，保留天然野趣；低干扰、高利用，保留、提升现状田园、荷塘、林木、土路，依托硬质场地改造设计活动场地。项目针对未来城市建设可能造成的径流污染，配

合水生态系统设置水下森林净化水体，助力保障漳江Ⅲ类水质，维护下游漳江口红树林生态环境；保护天然湿地海绵体，满足防洪排涝要求，协调规划竖向标高，保证堤外场地的安全性。

区域统筹，构建蓝绿交织、互联互通的生态休闲区

项目所在地西北承接云霄县城生活区，南部连接省级重点开发区——云霄县云陵工业开发区，东部下游方向是佳洲岛、漳江红树林保护区，是云霄县“三生”发展格局的重要节点，城、水、山在此交汇。设计师从统筹发展的角度出发，强调以水为脉、互联互通，把风景融入日常生活，借助润城之水，与下游湿地共同打造“商、养、学、闲、情、奇”六位一体的生态发展区，塑造功能复合的活力引擎，助力协调云霄县城“三生”发展格局；以南湖项目为节点，统筹建设文化和绿色基础设施，完善水系、绿道网络，打造县城郊野绿道环，链接云霄其他旅游、文化资源，包括开漳故城、威惠庙、将军山公园（陈政墓园）等，激发空间活力。

轻资产运营，构建宜居宜游的城市活力核

城市文化方面，云霄是漳州文明的发祥地，唐朝初年，陈政、陈元光戍闽开漳，实践儒家政教思想，传播中原文化（包括农耕文化），影响遍及漳、泉、潮、汕诸州及台湾地区，饮誉海内外；产业发展方面，云霄历史上以农业为主，在漳南居举足轻重的重要地位，近年来光电产业发展迅速，“光电之都”成为城市新的发展定位。

南湖生态园的定位为湿地海绵体，同时也是服务于周边的城市公园。设计师以文化和产业为切入点，改造现状建筑场地，结合出入口，布局主要服务建筑及场地，并选取优质景观资源点，布局小型多功能场地，通过轻资产运营的方式，承载多元活动，将市民休闲与旅游体验完美融合。推动社区营造，营造有归属感的宜居环境，为周边居民服务；接轨国际标准，打造国际营地综合体，吸引国内外游客，塑造云霄南湖旅游品牌。

项目建设了集水、田、林、村于一体的五感体验自然教育营地，开展自然课堂，方便孩子们学习水利、水生态、田林知识，了解不同生境的动植物的多样性，在大自然的探索中成长；依托“二十四节气”非物质文化遗产，建设闽南农耕文化、开漳文化的传习基地，开展农事体验、美食庆典、田园艺术、民俗表演等活动，树立“云霄寻根之旅”活动新品牌；推广科学健身理念，结合绿道及多功能场地，布局点线结合的健身系统，开展田林健身、智慧健身，将科学化、专业化、品牌化的健身项目和服务引入其中，打造户外运动营地；引导光电企业共建智能设施，增加人与自然的互动，并结合主题活动、庆典策划，联动周边区域，扩大整体旅游影响力。

城市的迅速扩张使城市湿地环境压力日渐加大，如何在城市稳定发展的基础上保护利用城市湿地，成为当前亟需解决的问题。云霄蒲美、南湖润城、南湖湿地作为漳海生态旅游带上的多元环境交汇点，其景观营造将是一项长期的系统工程。因此，必须坚持可持续发展观，维护湿地生态功能的完整性，防止湿地生物多样性退化，将生态保护与适度利用有效结合，并通过科普教育、文化体验、休闲观光等公众参与项目，实现湿地整体环境效益、社会效益、经济效益的最大化，为漳州“田园都市、生态之城”城市建设作出贡献。

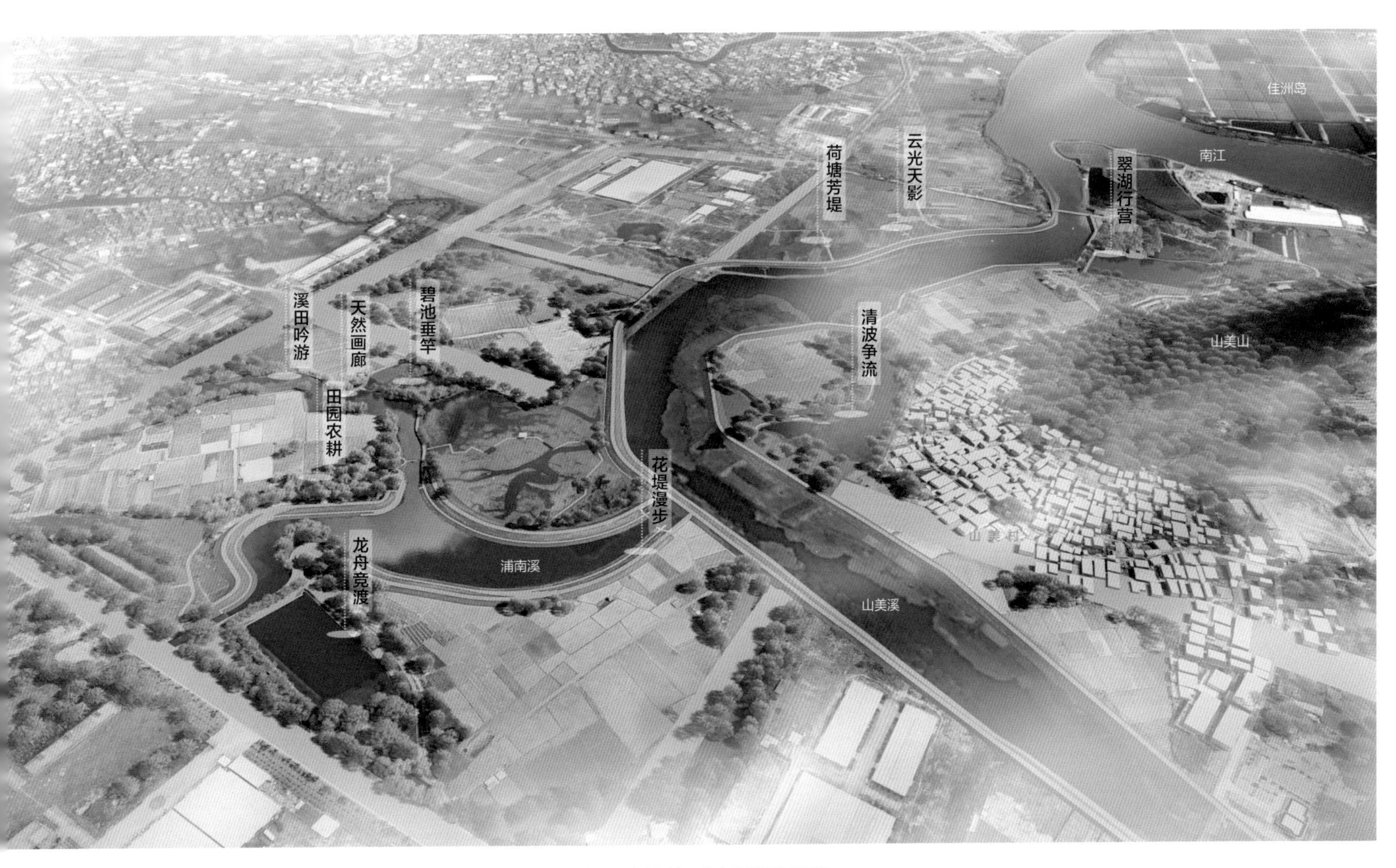

紧密结合现状及规划条件，着力保留天然野趣

Closely combining existing conditions and planning requirements, emphasizing the preservation of natural charm

改造现状堤岸，塑造湿地景观

Transforming the existing embankment to create a wetland landscape

梳理现状资源，利用当地材料

Utilizing local materials by organizing existing resources

场地现状

治理现状荷塘，提升观景设施
Rehabilitating existing lotus ponds and enhancing viewing facilities

净化水体，增加旅游休闲设施

Purifying the water bodies and adding tourist and leisure facilities

梳理驳岸，改造现状建筑为旅游服务设施

Organizing the waterfront and transforming existing buildings into tourist service facilities

建成实景照片　Real-life photos of the completed project

建成实景照片　Real-life photos of the completed project

建成实景照片　Real-life photos of the completed project

06 漳州市华安县真武山公园设计

Landscape Design of Hua'an County Zhenwu Mountain Park

- **项目地点：** 中国 漳州
- **项目规模：** 0.369 平方千米
- **设计时间：** 2016 年
- **施工时间：** 2017-2018 年

- **Project location:** Zhangzhou City, China
- **Project scale:** 0.369 square kilometers
- **Design period:** 2016
- **Construction period:** 2017-2018

景观设计着重发挥场地竹林资源特色，选择对自然干涉最小的生态措施和因地制宜的设计手法，改善公园环境，拓展公园多种综合功能，成为厦漳泉地区的旅游休闲目的地。

The landscape design emphasizes the unique bamboo forest resources of the site, employing ecological measures with minimal interference with nature and site-specific design techniques. The goal is to improve the park environment and expand its diverse functions, making it a popular tourist and recreational destination in the Xiamen-Zhangzhou-Quanzhou region.

总平面图

Master plan

真武山山地公园位于福建省漳州市华安县城北侧，真武山南麓，项目总用地面积约0.369平方千米。公园前身为创建于1992年的华安竹类植物园，园区地形主要由山坡和山麓组成，蜿蜒起伏，相对高差200米左右，形成了竹林、水库、山溪、谷地等特色景观资源，绿化覆盖率极高，被称为华安县的后花园，是久居城市的人们回归山林、享受自然的理想场所。

针对场地现状问题，构建生态修复系统

现状地被缺失、黄土裸露问题严重，存在水土流失等生态隐患，景观效果不佳。设计以生态、经济的设计手法，选用块石、竹子等当地材料和乡土植物，通过工程措施与植物种植相结合的方式，修复破损山体，构建经济集约的生态修复系统。

针对修复面积较大、危险等级较低的低土坡、土坎区域，选用当地竹材、自然石块为固坡材料，采用竹桩、竹垄、干垒石块为护坡形式，有效缓解土壤裸露边坡等水土流失问题，营造富有山林野趣的景观效果；针对高差大、土石不稳定、有较大的安全隐患的高边坡区域，采用石笼挡墙护坡，建造时在缝隙中插入活的植物枝条，覆盖人工痕迹；针对无需进行固土的边坡，由于表面植物生长困难，采用施工工艺简单，迅速见效的生态草毯建立垂直绿化。

植物修复依托本地的植物资源优势，适地适树，营造多样的植物景观类型，着重发挥植物护坡固土、水土保持、多样性营造等生态功能。在高边坡处结合工程修复措施，选择多年生藤本、蔓生或攀缘植物，弱化人工痕迹，绿化山体，同时减缓降雨对土壤的直接冲刷，保持水土。针对竹林区，通过增加现状竹林的伴生树种，补植各类观花植物、观果植物，吸引昆虫、鸟类、爬行类甚至小型哺乳动物，形成较为完善的生态链，提高物种多样性，提升山体动植物生境。

利用现状道路，营造特色游赏路径

公园目前的道路多为断头路，缺乏完整的园路体系。设计在现有园路基础上结合山体环境特征，以较少的投入，完善游步道体系，形成三条特色游赏路径：①利用现状柏油路设置特色健身径，全程4.5千米，满足骑行、慢跑、徒步等健身运动；②整合现状高差变化大的登山小径，形成以竹林养生、竹林科普、竹林休闲、竹林艺术为主题的“竹林浴”体验径。线路全长1.5千米，具有观景、远眺、休憩、运动等多种使用功能；③围绕真武山水库、溪涧等自然景观设置综合体验径，全长1.5千米，沿路设置饮食、养生、康复、民族风情、动植物科普等功能设施，突出五感体验，在湖边、谷地等亲水区域营造自然、舒适的健身及游赏环境，让游客得到放松身心、寓教于乐的游山环湖感受。

发掘地域优势资源，彰显地方文化特质

华安四季温暖多雨，竹源丰富，目前已从全国各地引种30属330种竹种，是国内种植面积最大、竹类品种最多、属性最全、功能最齐的竹子基地之一，也是竹类科研教学、竹子生产推广、旅游观光休闲、学术交流示范的综合性基地。

景观充分利用竹林资源优势，设置瑜伽平台、休闲座椅、趣味吊床等休憩设施，并提供辅助标识信息指导健身、放松，让人们体验真武山独具特色的竹林浴。项目还邀请艺术家、院校学生在竹林中创作竹墙、竹笼等竹类艺术装置设施，突出竹文化主题。设计进入汇水谷地和两侧竹林的游赏步径，形成竹溪游赏空间。

除华安竹外，华安玉、华安茶文化也在公园也有所体现。公园南入口采用福建省奇石品种、漳州市市石“华安玉”设计枯山水景观。西入口设计观景平台俯视茶田风光。园内设置品尝“华安茶”，体验茶文化，进行茶道学习的场所。

作为自然环境优良的依托山林基底的城市公园，设计尊重自然环境，以对自然生态影响最低为前提，使公园具有绿道串联、休闲健身、家庭亲子、科普养生、旅游观光等多种综合功能，最大化的发挥社会、生态、经济等综合效益。

依托现状竹林资源突出竹文化主题　Highlighting bamboo culture by leveraging existing bamboo forest resources

改造现有柏油路为多功能的健身绿道　Transforming the existing asphalt road into a multi-functional fitness greenway

与山水相融的涉溪步道　Streamside trails harmoniously integrated with the surrounding landscape

户外廉政教育基地“亮节讲坛” Outdoor integrity education base

户外茶席 Outdoor tea seating area

山顶观景台　Mountain top viewing platform

华安城景观赏点

Hu'an City scenic viewpoint

依托现状水库建设的亲水空间

Waterfront space created around the existing reservoir

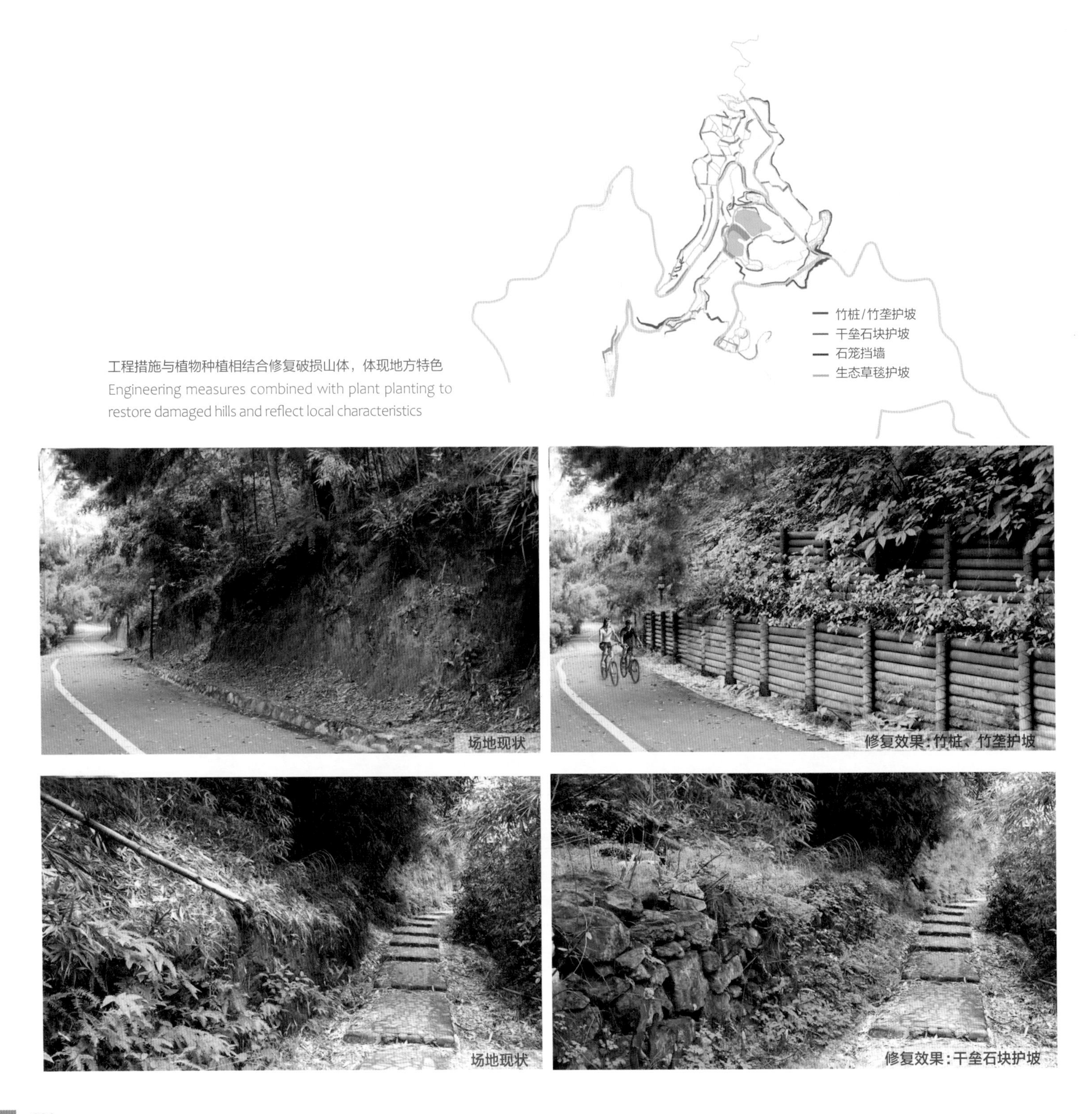

工程措施与植物种植相结合修复破损山体，体现地方特色

Engineering measures combined with plant planting to restore damaged hills and reflect local characteristics

场地现状
修复效果：石笼挡墙护坡
场地现状
修复效果：生态草毯护坡（高边坡）
场地现状
修复效果：生态草毯护坡（硬化堤岸）

建成后的实景照片　Real-life photos of the completed project

建成后的实景照片　Real-life photos of the completed project

绿道
统筹建设

COMPREHENSIVE CONSTRUCTION OF GREENWAYS

绿道是以自然要素为依托和构成基础，串联城乡绿色开敞空间，以游憩、健身为主，兼具市民绿色出行和生物迁徙等功能的廊道。绿道作为一种功能复合的线性土地网络，是实现统筹建设的重要载体之一。

Greenways are corridors that connect urban and rural green open spaces, relying on and incorporating natural elements. They primarily serve recreational and fitness purposes, while also facilitating green transportation for citizens and supporting wildlife migration. As a multifunctional linear land network, greenways play a crucial role in comprehensive construction.

厦门市海沧区与厦门本岛隔海相望，是厦门岛外发展的城市副中心、国家级台商投资区。产业的蓬勃发展吸引了大量外来人口，城镇化发展迅猛，现有城市绿地难以满足市民日益增长的户外休闲需求。我们自2012年以来陆续承接了海沧区若干重要绿色公共空间的景观规划设计工作，将绿道作为“城市双修”的辅助框架，有机连接分散的生态斑块，强化生态联通和“海绵”功能，参与构建区域性生态网络；为市民提供开放共享的绿色休闲健身场所，丰富城市绿色出行方式，促进厦门岛内外协调发展。

厦门市海沧区绿道统筹建设项目是大型动态复合城市景观项目，先期完成景观系统的整体策划及定位，然后分区逐步深化进行景观设计。我们将绿道和城市公共景观作为绿色基础设施，目的是建立一个能够整合各种资源、使城市一体化发展的平台。景观设计将在解决生态问题的同时延续地域文化，带动经济发展，引导公众参与城市建设，培养市民健康积极的生活方式。

Located across the sea from Xiamen Island, Haicang District is a secondary urban center and a nationally designated area for Taiwanese investment, experiencing rapid urbanization and attracting a significant influx of migrants due to its thriving industries. The existing urban green spaces are unable to meet the growing demand for outdoor recreational activities. Since 2012, we have been involved in the landscape planning and design of several important green public spaces in Haicang District. Greenways have been established as an auxiliary framework for " dual urban cultivation," connecting dispersed ecological patches, enhancing ecological connectivity and "sponge" functions, and participating in the construction of a regional ecological network. These greenways provide open and shared green leisure and fitness spaces for citizens, enrich urban green transportation options, and promote coordinated development between Xiamen Island and its surrounding areas.

The Comprehensive Construction of Greenways in Haicang District, Xiamen, is a large-scale dynamic composite urban landscape project. The landscape system's overall planning and positioning have been completed in advance, followed by progressive deepening and landscape design in different zones. We consider greenways and urban public landscapes as green infrastructure, aiming to establish a platform that integrates various resources and facilitates integrated urban development. The landscape design not only addresses ecological issues but also perpetuates regional culture, drives economic development, encourages public participation in urban development, and cultivates a healthy and active lifestyle for citizens.

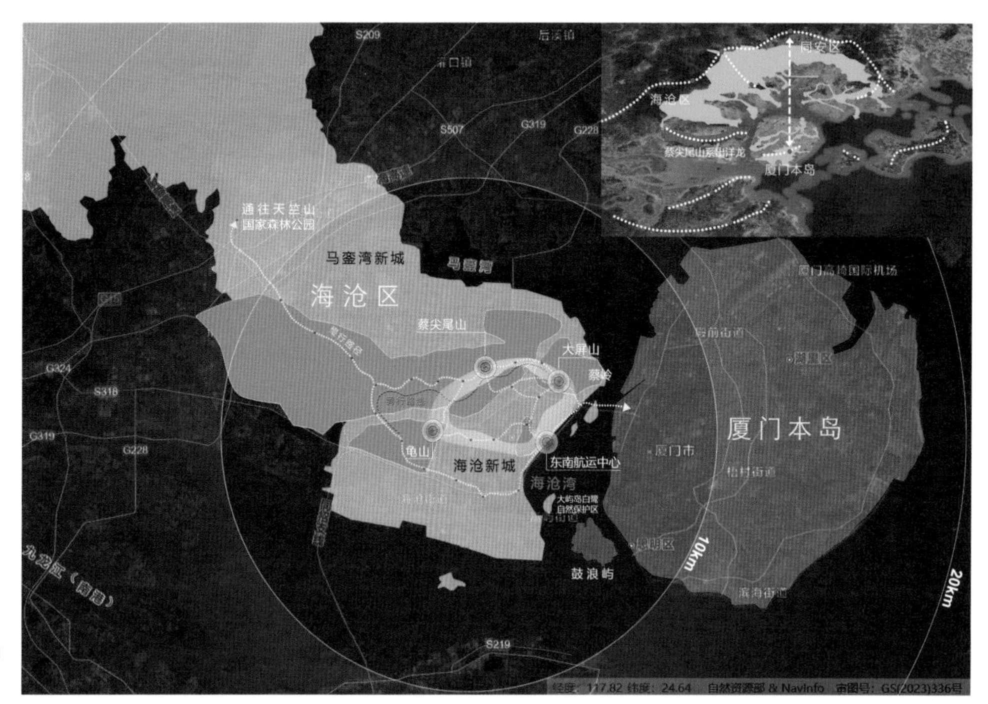

通过绿道对海沧区的绿色开放空间系统进行整合

Integrating the green open space system in Haicang District through greenways

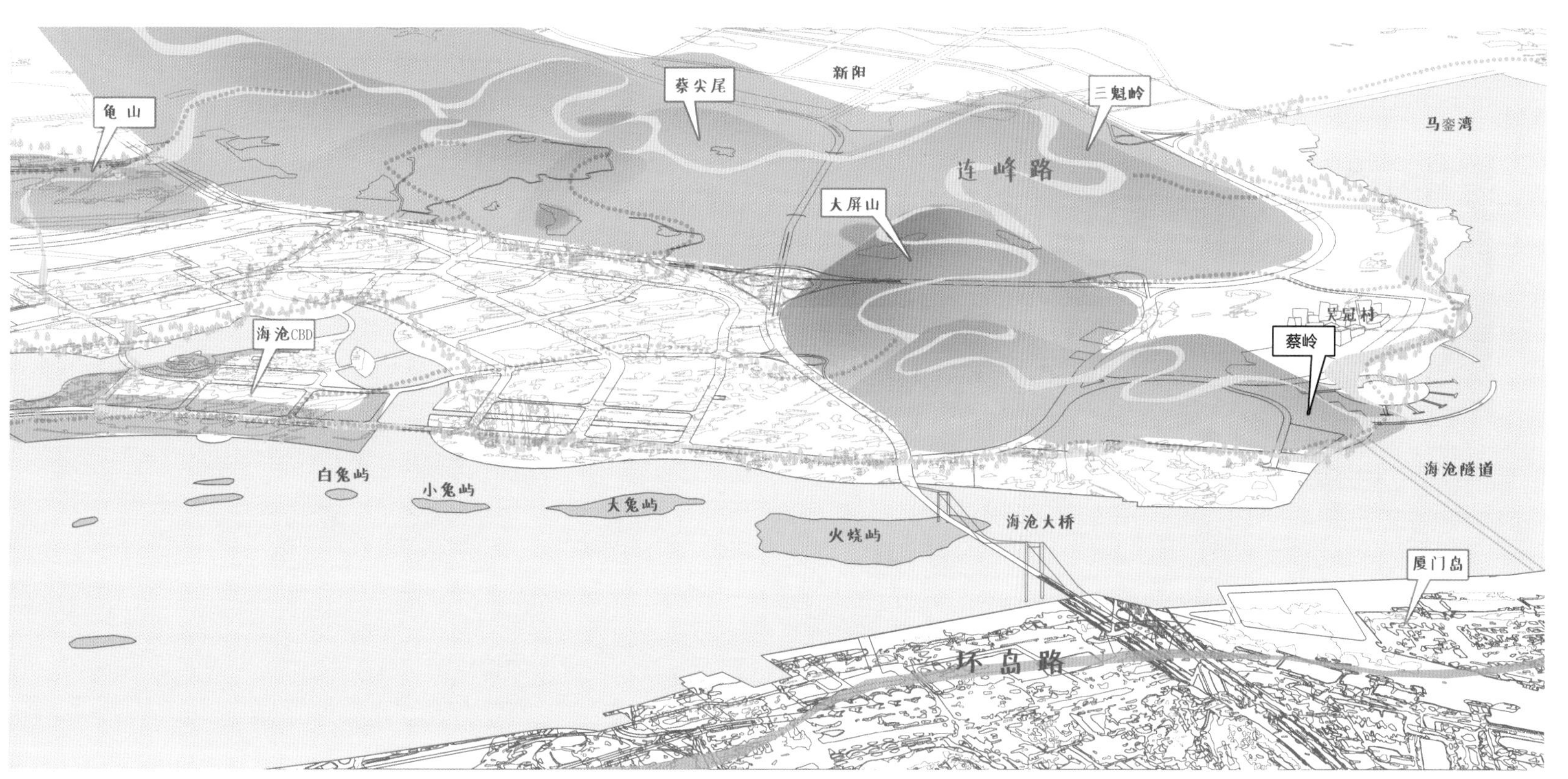

01 厦门蔡尖尾山、大屏山绿道规划设计：差异化发展的城市绿廊

Planning and Design of Xiamen Caijianwei Mountain and Daping Mountain Greenway: Differentiated Development of Urban Green Corridors

- **项目地点：** 中国 厦门
- **项目规模：** 约 3.30 平方千米
- **设计时间：** 2011-2013 年
- **施工时间：** 2013-2017 年

- **Project location:** Xiamen, China
- **Project scale:** 3.30 square kilometers
- **Design period:** 2011-2013
- **Construction period:** 2013-2017

蔡尖尾山与大屏山是海沧两大山系，绿道建设与山体修复、林相改造等工程结合，保护修复山地海绵体，在蔡尖尾山布局多样化的森林健身场地，在大屏山营造以山林公园为形象的门户地标和新兴旅游目的地。

Caijianwei Mountain and Daping Mountain are two major mountain ranges in Haicang District. The greenway construction integrates with mountain restoration, forest transformation, and other projects to protect and restore the mountainous sponge, establish diverse forest fitness areas in Caijianwei Mountain, and create a gateway landmark and emerging tourist destination with a mountain and forest park image in Daping Mountain.

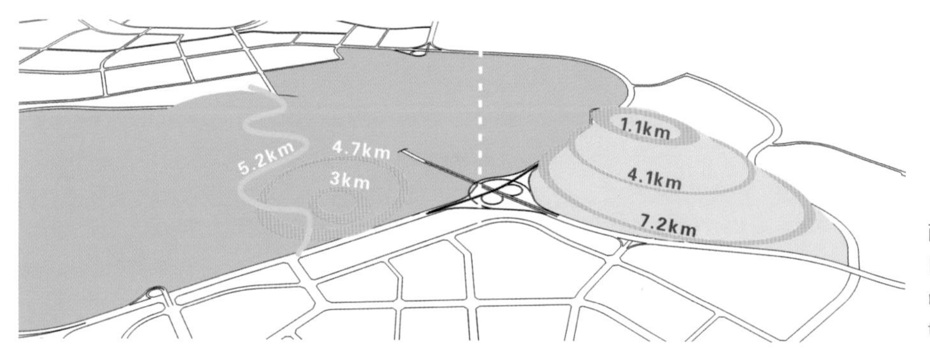

两山差异化发展定位
Differentiated development positioning for the two mountains

蔡尖尾山和大屏山位于海沧区中南部，是海沧区“两山两湾两城”城市格局的重要部分——区域绿心。项目周边规划地块主要为居住用地、商业金融业用地、行政办公用地、文化娱乐用地等。

蔡尖尾山森林茂密，现状有水库、水塘等自然汇水系统，水库边环境较为优良，存在多处废弃采石场，承载了地区的特殊历史记忆。大屏山过火林区域补植后植被生长不良，尚未形成密林，山上以无组织排水为主，水流冲刷地面形成沟壑，修建的路侧排水沟尚未完成。两山土质较为瘠薄，易水土流失，且因采石、道路修建等造成部分岩石裸露，岩石表面浮土现象严重。两山鸟类及昆虫种类丰富，生态系统较为完善。

依山就势，山体绿道设计充分结合现状

蔡尖尾山和大屏山现状已建成部分车行路和人行道，但道路不成系统，亟待提升完善。设计充分利用现状资源，在保证设计合理的前提下实现对现状的最小干预。

车行路设计利用现状地势，对于视野开阔、空间开敞的区域，设计观景挑台、观景座椅等游憩设施，留出观景视线通廊，俯瞰城市，同时沿道路及水岸补植遮阴及观赏植物，营造舒适的空间环境；对于道路两侧植被茂密，空间围合度高的区域，结合山形地势，在道路两侧设计艺术化座椅及休息场地，同时沿道路及两侧坡地补植观赏植物，打造富有自然野趣的山林绿道。

人行路分亲水型和山野型两类，亲水型人行路紧邻水体，视野开阔，沿水岸设计亲水栈道、亲水平台、游憩设施和林下休息场地，沿栈道两侧及水岸补植景观植物，将线路隐藏于树木花草之中，丰富游赏体验。山野型人行路坡度较大，两侧植被茂密，空间围合度高，在视野开阔处设计观景挑台、休息座椅等游憩设施，局部补植凤凰木、火焰木等观花乔木和庭荫乔木，突出特色，显现山野情趣。

蔡尖尾山和大屏山现状道路两侧存在生态破坏和环境隐患，主要表现为道路两侧坡度陡峭，土壤裸露；部分岩石裸露松动；水土流失；无组织排水等。设计有针对性地提出了山体修复策略和水体修复策略，采用垂直绿化、绿化防护、设置截水沟、削坡和圬工护坡等手段，对重要的生态环境问题加以修复。

两山统筹发展，差异化定位，一动一静，相辅相成

蔡尖尾山规划主题为享受宁静。蔡尖尾山现状山林茂密，有唐代古寺石室禅院及明代古寺石峰岩寺等文化资源，适于进行禅修及森林浴活动。禅修活动与森林浴的结合具有独创性，从内容上满足人们对身、心两方面进行调节的需求，形成完善的养生健身系统，从设施上两者的要求近似，可以共同使用，提高设施的使用效率，节省投资造价。

大屏山规划主题为活力地标。大屏山紧邻海沧大桥，地理位置优越，可俯瞰厦门全景，拥有观看厦门与大海的独一无二的视角，是吸引游人的主题之一。厦门开放的经济、文化基础，以及大屏山周边创意产业园区等均有利于将大屏山变成厦门创意活动的发布地，一系列的创意活动都将在厦门城市背景下展开，引领全国乃至国际性创意活动的潮流。

蔡尖尾山和大屏山拥有丰富的自然资源，通过自然手作坊、手工市集、环保艺术节等多种活动，鼓励市民到山间学习大自然的智慧，将环保、回收的意识带进生活之中。

蔡尖尾山、大屏山绿道于2017年陆续建成开放，目前已经成为海沧市民休闲、游憩、健身的重要场所，也成为游客重要的旅游目的地之一。

营造以山林公园为形象的门户地标

Creating a gateway landmark with a mountain forest park as its image

站在大屏山公园西区山顶新建的厦门岛外观景台上可以一览厦门全貌

Building a new viewpoint on the western peak of Daping Mountain in Daping Mountain Park which is outside Xiamen Island offering a panoramic view of Xiamen

大屏山山地景观　Mountain landscape in Daping Mountain

大屏山公园东区结合现状地形设计8千米长的无障碍健身径，提供全年龄休闲活动场所

The eastern section of Daping Mountain Park integrates the existing topography and features an 8-kilometers-long accessible fitness trail, creating a year-round recreational space suitable for people of all ages

蔡尖尾山修复了山体植被，将森林防火道与登山健身道一体化设计

Restoring the vegetation on Caijianwei Mountain, integrating forest fire prevention and mountain hiking trails

结合现状水库设计的禅修场所
Designing a zen meditation place combining nature and human elements by using the existing reservoir

蔡尖尾山建成后的实景照片
Real-life photos of the completed project on Caijianwei Mountain

蔡尖尾山建成后的实景照片
Real-life photos of the completed project on Caijianwei Mountain

02 厦门中心公共空间景观设计：慢生活 CBD 体验式绿道

Landscape Design of Xiamen Center Public Space Project: Experience-oriented Greenway for Slow Living in CBD

- **项目地点：** 中国 厦门
- **项目规模：** 0.13 平方千米
- **设计时间：** 2013-2018 年
- **施工时间：** 2015-2019 年

- **Project location:** Xiamen, China
- **Project scale:** 0.13 square kilometers
- **Design period:** 2013-2018
- **Construction period:** 2015-2019

厦门中心公共空间生态系统从一体化统筹的理念出发，将湖景、海景、市景的生态环境相互串联，使生态环境与城市环境有机共生。针对不同区域、不同功能、不同空间构建相适应、可持续的生态环境系统。

The ecological system of Xiamen Central Public Spaces is designed based on the concept of integrated planning, connecting the ecological environments of lakes, seas, and the cityscape, fostering a symbiotic relationship between the natural and urban environments. Adapted and sustainable ecological environments are constructed to suit different regions, functions, and spatial requirements.

鸟瞰效果图 A bird's-eye view rendering

一体化统筹生态系统

项目场地位于湖海之间，视野开阔，对望厦门本岛和“鼓浪屿”，东南1千米处有“白鹭保护区”大屿岛，场地内建筑群是厦门最大的商务综合体，涵盖写字楼、购物中心、豪华酒店、会展中心、厦门水秀（水秀公园）等全方位业态。

海沧湾位于场地东侧，原以滩涂为主。沿海向北是一座小型避风港，四周有一片原生红树林。政府计划对海沧湾进行清淤以增加纳潮量，增强海水自净能力，提升水域内的生态环境，同时启动整体岸线建设和周边岛屿的综合开发和保护。地块西侧湖岸边建设了大量的石材矮墙，因场地内地面高低不平，矮墙也随之时高时低，面向湖面的视线受到一定的遮挡，水边是硬化的驳岸和几何型的亲水平台，构成呆板的现状湖岸。

在此基础上，景观设计需对地块内以及地块周边统筹提升，既要联系城市，又要整合湖海资源，形成整体的城市景观以及连续的湖海生态系统。

从城市发展角度进行整体考量，既要展现城市各界面的景观面貌，同时也必须结合场地内外的功能、形式、业态等诸多因素，在保持视线通透性的同时提升片区的绿化率与绿视率。地块内有城市绿道贯穿延续至滨海，形成一条联系城市南北向的绿色廊道。结合保留原生红树林的同时对现有红树林进行增补，形成海上植物园，与大屿、大兔屿、小兔屿等周边岛屿的白鹭生态保护区形成连接。

在整体植物品种的选择上尽量选用适应性强的乡土植物，便于后期管理以节约成本。在植物的搭配上，注重考虑地域性特征，以及与现状建筑使用功能、空间特点的协调性，运用精致细腻的植物搭配手法，营造具有亚热带海洋风情的植物空间。

沿海一侧种植在考虑抗风性的同时，结合区域功能和游人的使用特点，以开敞、半开敞空间为主。靠近建筑入口以及靠近海沧大道的部分，上层以王棕、海枣等棕榈科植物相互搭配为主，展现开敞空间并突出海洋风情，其余部分选用香樟、高山榕、秋枫等高大遮阴乔木搭配中层及下层棕榈科植物营造精致的开敞、半开敞空间。植物生态桥与花园式的下沉广场植物搭配细腻精致，形成与独特海岸风貌的生态联系。

湖岸一侧进行系统化提升，充分考虑从建筑底层将视线延伸至湖岸的策略，形成通透而自然的亲水坡岸。利用沿湖一侧建筑基底到湖岸线的高差形成三层具有商业休闲功能的台地花园，紧邻高大建筑正面一侧种植王棕，建筑侧面种植相对低矮的狐尾棕，在纵向上与建筑相互穿插，形成丰富的视觉层次，同时也满足视线上的通透要求。坡上具有热带风情的灌木地被组合搭配鸡蛋花，形成一个个组团群落。原有几何形状的硬质平台改造成边界变化、空间丰富的亲水花园，亲水花园结合微地形种植鸡蛋花、海枣、旅人蕉、散尾葵、龟背竹、鹤望兰等植物组团，贴近水岸则选择耐盐碱植物共同组成细腻精致的植物花园。中心商业广场四周由绿岛围合，在保证商业活动的同时，形成一大片舒适的林下空间，与广场形成功能上的互补。酒店花园相对独立，复层种植结构形成幽静且富有场所感的花园景观。场地中大量的复层种植构成各式各样舒适宜人的室外空间，鼓励办公人群及市民充分利用室外空间，尽量减少室内活动时间从而节约更多能耗。

拆除湖岸边阻隔视线的实体矮墙，通透的金属栏杆使沿湖一侧的视线全部打开。场地“见缝插绿”，形成了数量众多的绿岛花园，建筑屋顶的退台种植大量植物，形成层叠的屋顶花园，提高了整个场所的绿地率与绿化覆盖率。微地形结合丰富的植物群落，既优化了小气候环境，也提升了绿视率。

场地内大面积采用透水铺装增加下渗性，线性收水沟、导水槽及排水沟将未下渗的雨水收集至蓄水模块中，统一净化，回收利用。不同地形选择相应的灌溉系统以节约水资源。

综合措施优化了厦门中心的公共空间以及周边的生态系统，丰富了生物多样性，同时构建了更加舒适的小气候环境，建立了白鹭保护区以及周边与场地的生态廊道，完善了地块内部与周边环境的整体生态系统。

建成后的实景照片

Real-life photos of the completed project

建成后的实景照片
Real-life photos of the completed project

03 厦门龟山公园绿道规划设计：非遗民俗活动保护与传承的活力路线

Planning and Design of Xiamen Guishan Park Greenway: Vibrant Route for the Protection and Inheritance of Intangible Cultural Activities

- **项目地点：**中国 厦门
- **项目规模：**0.726 平方千米
- **设计时间：**2015 年

- **Project location:** Xiamen, China
- **Project scale:** 0.726 square kilometer
- **Design period:** 2015

设计采用渐进式有机更新，构建区域健康生态绿核，联动厦门绿道网络，承载民俗活动，活化传承非物质文化遗产，促进回迁原住民与海沧新移民的融合，创建宜居新区。

Utilizing progressive and organic renewal approaches, the design aims to establish a regional healthy ecological green core, integrate with Xiamen's greenway network, accommodate folk activities, revitalize the intangible cultural heritage, promote the integration of original residents and migrants in Haicang, and create a livable new district.

龟山公园是规划临港新城的中心绿地，是回迁原住民与海沧新移民共享的公园

Guishan Park is the central green space in the planned Lingang New City, shared by both relocated indigenous people and new immigrants in Haicang District

龟山公园位于厦门市海沧区临港新区中心，北靠海沧区两大生态绿核之一的蔡尖尾山系，南邻海沧港区，既是临港新区的后花园，也是蔡尖尾山生态绿核向南延伸到海沧港区形成的山海通廊的中心环节。龟山公园规划面积约0.726平方千米，是临港新区在快速城镇化进程中唯一保留的生态斑块。场地现状资源如何利用，地域文脉特色如何延续，与区域发展规划如何衔接，是龟山公园景观规划设计需要解决的核心问题。

尊重场地现状，秉持渐进式更新，塑造淳朴野趣的自然基底

龟山公园场地现状地形特点为东西两山并立，场地东侧的网山为公园制高点，也是区域的视觉焦点；东西两山之间坡度较为平缓，呈马鞍形，适宜开展需要一定场地的群体性活动。现状土地以林地为主，覆盖率达52.3%，主要种植龙眼林；场地边缘分布有零星池塘及洼地，水质较差。

景观设计尊重场地现有的自然地理脉络，依山就势，尽可能地降低对自然山体的人为影响，保留长势良好的现状树林和具有乡土特色的农田肌理，维护场地及周边原有的生态格局。注重现状资源的保护与利用，保留场地中祠堂、庙宇等历史价值较高的地文地物，并使其融入新的公园环境；提取乡土建筑设计语汇，重复利用拆迁建筑材料，将其作为特色景观元素在龟山公园中加以传承与发展；保留现状三处水体成为公园的汇水区，形成四季水景。整体设计坚持可持续发展策略，摒弃跃进式更新，采用渐进式有机更新，节约投入，保证公园在不同建设时期的景观效果及使用功能。

单一的植物群落不利于生物多样性，难以形成舒适的公园小气候，也不能满足公园多样化景观的需求。设计将现状大面积的龙眼林划分为中心区和边缘区，采取不同的更新策略：中心区域生态敏感度高，破坏后极难恢复，以现状保护为主；边缘区生态敏感度低，以补种替换弱小树、濒死树为主，同时根据公园在使用功能和景观效果等方面的不同需求，进行多样化的更新补植，经过植物群落的逐步演替，与保留的龙眼林共同形成稳定的人工植物群落，营造舒适宜人的环境，构建区域健康生态绿核。联动厦门绿道网络，串联两山、两湾、海沧、本岛，逐步扩大区域影响力，合力构成海沧区环境新名片，创建宜居新区。

挖掘地域文脉，彰显海沧特色，营造蕴含传统文化的时代场所

龟山公园西北部紧邻的全国重点文物保护单位青礁慈济东宫，是为纪念保生大帝吴夲而建立的庙宇。保生大帝被民间谥为“医灵真人”，保生大帝的民间信俗广泛分布于福建西南部、广东东部、台湾地区及东南亚各国华人地区。海沧青礁慈济东宫自2006年始，已成功举办12届海峡两岸（厦门海沧）保生慈济文化旅游节，每年吸引海峡两岸以及东南亚的上千名各界人士，已成为两岸交流联谊的重要品牌。公园以青礁慈济东宫的保生大帝雕像为视觉统领，联系中心活动广场、南入口，从视线上将龟山公园与青礁慈济东宫整合联系起来。以改造拆迁留下的空地为活动场地，策划举办游境踩街、青草药课堂、民间艺术节和两岸汉字文化交流等活动，打造“共同信俗、共同语言”的示范性主题公园，使龟山公园成为两岸文化交流会场，扩大公园影响力，强化公园品牌。

策划先行，为传统游境踩街活动设计路线，将龟山公园、周边社区与青礁慈济东宫整合联系起来

Through the proactive planning of a route for traditional parade activities, the Guishan Park, surrounding communities, and Qingjiao Ciji Palace are integrated and connected together

将现状洼地水体进行生态化、景观化改造，营造多样的城市界面，解决雨洪问题
Ecological and landscape transformation of the existing low-lying water bodies to create diverse urban interfaces and solve problems related to rainwater flooding

打造层次丰富、乡土自然的公园入口形象
Creating a rich and indigenous park entrance image

保留村落祠堂，延续场所记忆

Preserving village temples to continue the sense of place

使用传统材料延续场地基因

Incorporating traditional materials to preserve the site's heritage

04 厦门蔡岭公园绿道规划设计：自然探索游线
Planning and Design of Xiamen Cailing Park Greenway: Nature Exploration Trail

- **项目地点：** 中国 厦门
- **项目规模：** 0.70 平方千米
- **设计时间：** 2015 年

- **Project location:** Xiamen, China
- **Project scale:** 0.70 square kilometers
- **Design period:** 2015

蔡岭位于厦门海沧区最东端，因其优越的地理位置和厦门旅游发展要求，设计定位为以山体及植被保护为主的综合性山体公园，为厦门市民及游客提供健身、休闲的新去处。

Located at the eastern end of Haicang District of Xiamen, Cailing Mountain is positioned as a comprehensive mountain park primarily focused on mountain and vegetation conservation. It provides a new destination for fitness and leisure activities for Xiamen's residents and visitors due to its excellent geographical location and the requirements for tourism development in Xiamen.

区域联系
Regional connections

先行制定发展策略，精细化主题定位

我们根据上位规划和现场调研情况提出了蔡岭公园的发展策略，并在此基础上进行详细的方案设计，将策划与设计紧密结合，确保项目落地后可以发挥生态、社会与经济效益。具体发展策略包括：从区域发展角度，提出以绿道串联的方式构建生态与交通双重廊道，形成工业区和居住区之间的绿色走廊，与吴冠村、大屏山、三魁岭连片发展；从保护性开发角度，提出绿道建设应结合现状登山径及山林防火道，局部进行生态修复，丰富植物种类，增加观山看海的观景休闲设施，完善标识、厕所等服务设施；从策划运营角度，将亲子健身与自然教育结合，成为户外家庭活动的举办地，为人们留下美好的公园记忆。此外，我们建议在大屏山和蔡岭之间架设景观步行桥，形成两山架一桥的海沧新大门。

山体修复、游线串联，塑造自然探索空间

蔡岭山体植被生长良好，林下空间适合进行各类活动。现状地形主要划分为山谷、山顶、山坡、山脚四类，山谷有汇水形成的水塘，植被茂盛，风景优美。设计对山体进行修复和雨水管理后，可以形成连续的生态水系，是自然群落丰富，可以进行各类观察、探索活动的场所和孩子们的自然课堂。山顶设置几处观景平台，可以眺望马銮湾和厦门本岛，成为厦门岛风光的观赏地。山脚设置公园出入口，面向海沧中学，成为开放的休闲活动广场。山坡设置登山步道、无障碍坡道和栈桥，串联场地，通过景观人行天桥与大屏山相连。步道设计选择经济耐用的本土天然材料，营造山林野趣的整体氛围。散布在游径上的设施和场地有利于开展各类交往活动，如趣味运动会、家庭野餐会、骑行之旅、望远镜眺望、青少年营地等，让孩子们在自然中健康快乐成长。

设计愿景
Design vision

总平面图　Master plan

A 自然创作区

1 北入口
2 镜湖（汇水湖）
3 池塘
4 绿溪
5 野餐厅
6 望景台

B 山地森林浴区

7 山林穿越
8 景观桥
9 树林探知
10 树林广场南入口
11 花径
12 树冠栈桥
13 闻香谷

C 自然认知区

14 彩虹坪
15 湿地花田
16 山谷讲学
17 草场地
18 岩石花园
19 浅滩
20 绿色峡谷
21 东入口

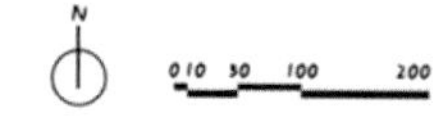

场地分成四种大的空间类型，通过无障碍系统将这四种空间类型进行串联，让公园成为各个年龄段都可以使用的共享空间

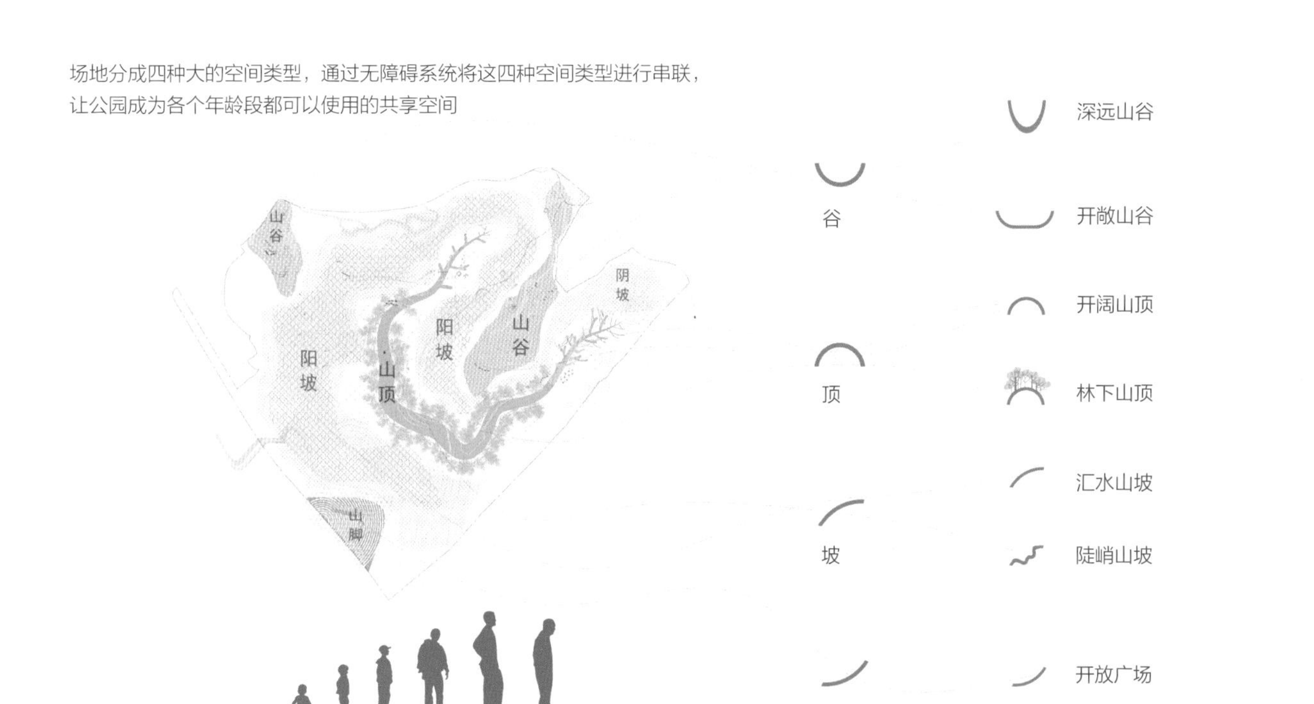

不同年龄段的所有使用者都可以进入公园的所有区域参与活动

无障碍系统串联4种山地类型
Accessible system connecting four types of mountainous areas

从山顶观景点观赏马銮湾是蔡岭独有的景观特色
Unique landscape feature of enjoying Maluan Bay from the mountaintop viewpoint in Cailing Mountain

山谷水池　Valley ponds

湿地花园 Wetland gardens

石头课堂 Stone classrooms

户外教育活动 Outdoor educational activities

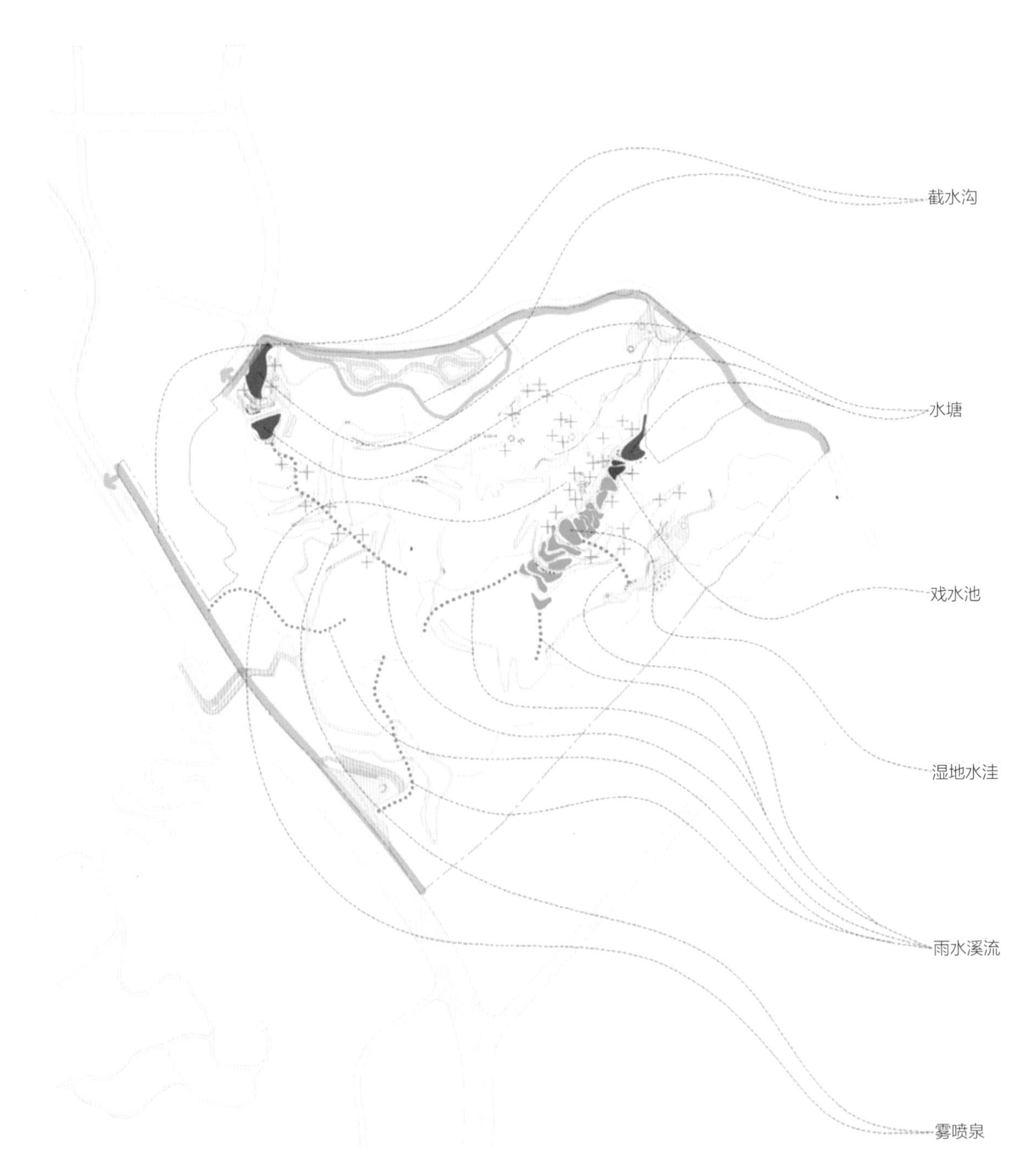

设计后，场地中的水会以截水沟、水塘、戏水池、湿地水洼、雨水溪流、雾喷泉六种形式呈现，多样的水景形式让使用者最大限度地参与到和水的互动中

多样的水景形式鼓励人们与水互动　Diverse water features encouraging interaction with water

绿色基础设施规划建设

COMPREHENSIVE CONSTRUCTION OF GREENWAYS

绿色基础设施是在灰色基础设施和社会基础设施两者基础上提出的生态规划概念，是在城市总体规划发展、生态保护战略上的一种新的规划方式。绿色基础设施规划将城市、乡村统筹考虑，利用城乡自然生态基底，如河流、山脉、林地、农田，再结合城市中公园、绿地等开敞空间，形成具有生态连续性和生态整体性的绿色网络，进一步提升人的生活环境和生活品质，并尽可能地恢复已经遭到破坏的生态系统。随着科技进步和生态意识加强，绿色基础设施规划正逐步与微观设施技术结合，更加具有可操作性和可实施性，如近年提出的“海绵城市”雨洪管理设施等。“海绵城市”是将城市比喻为海绵，城市下垫面具有海绵一样的吸、蓄水能力。下雨时城市下垫面可以吸水、蓄水、渗水、净水，补充地下水资源，在需要时，可以对这些地下水资源加以利用。

无界景观的相关项目实践旨在通过结合“海绵城市”建设，构建绿色基础设施框架，保证生态系统、城市基础设施、环境品质、资源循环、社会文化协调有序发展，突破传统生态保护的局限性，提升土地资源节约、集约利用水平，融入文化认同、公众参与和运营管理，实现生态、社会、经济的协调和可持续发展。

Green Infrastructure is an ecological planning concept that builds upon gray infrastructure and social infrastructure, offering a new planning approach for urban development and ecological conservation strategies. The planning of green infrastructure considers the coordination between urban and rural areas and utilizes the natural ecological foundations such as rivers, mountains, forests, and farmland, along with open spaces like parks and green areas in cities, to form a green network with ecological continuity and integrity. This approach aims to enhance the living environment and quality of life for humans while restoring damaged ecosystems as much as possible. With advancements in technology and increasing ecological awareness, green infrastructure planning is gradually integrating with micro-level facility technologies, making it more practical and implementable. Examples include the "sponge city" stormwater management facilities proposed in recent years. The "sponge city" concept compares the city to a sponge, with the urban underlying surface having the ability to absorb, store, infiltrate, and purify rainwater. This approach replenishes groundwater resources and utilizes them when needed.

The practical projects undertaken by View Unlimited Landscape Architects Studio aim to construct a green infrastructure framework by integrating the concept of "sponge city". This approach ensures the coordinated and orderly development of ecosystems, urban infrastructure, environmental quality, resource circulation, and social-cultural harmony. It goes beyond the limitations of traditional ecological conservation and enhances the level of efficient land resource utilization. The projects also incorporate cultural identity, public participation, and operational models to achieve coordinated and sustainable development among ecology, society, and the economy.

01 重庆两江新区悦来新城后河环境综合整治工程

Chongqing Liangjiang New Area Yuelai New City Houhe Environmental Remediation

- **项目地点：**中国 重庆
- **项目规模：**1.22 平方千米
- **设计时间：**2016-2017 年
- **施工时间：**2018—
- **合作单位：**哈尔滨工业大学

- **Project location:** Chongqing, China
- **Project scale:** 1.22 square kilometers
- **Design period:** 2016-2017
- **Construction period:** 2018-
- **Cooperating unit:** Harbin Institute of Technology

我们在此项目中提出以立体山地海绵系统为基础，探索具有示范意义的新城开发模式，以洁净、美丽、活力为目标，将海绵涵养廊道、生态景观廊道、活力休闲廊道合一，切实保护自然河谷生态本底，智慧应对新城开发。

In this project, we propose a three-dimensional mountainous sponge system as the foundation to explore a demonstrative new urban development model. The goal is to create a clean, beautiful, and vibrant environment by integrating sponge retention corridors, ecological landscape corridors and dynamic recreational corridors. The design aims to effectively protect the natural valley's ecological foundation and employ smart strategies to address new city development.

后河是嘉陵江左岸一级支流，是重庆山水体系的重要组成部分，是优质的天然海绵体，悦来新城是重庆两江新区西部片区的核心区域，设计区域位于悦来新城与水土、渝北两区的交界处、后河的最下游，后河在此地向西汇入嘉陵江，对嘉陵江水质有直接影响。

设计范围包括后河干流段、猪肠溪河段及猪肠溪支沟段。场地现状为纯天然河道，河谷横断面呈深U形，邻河山体局部坡度较大（大于30%），河道纵断面存在7处跌落。现状基本为Ⅲ类水质，存在一定面源污染。设计红线内河谷大部分为自然植被覆盖的坡地，有部分人工梯田及若干片弃土堆填区，生态敏感性较高。现状谷深弯多，多处制高点可远眺观景，亲水视线变化丰富。场地周边尚无城市建设，规划以科研、居住用地为主。

本项目涉及流域综合整治、海绵城市、山体修复、生态景观、市政综合管网、智能监控监测、轨道保护等内容。景观设计团队发挥统筹协调作用，与哈尔滨工业大学重点实验室、中国科学院水生生物研究所、中国科学院微生物研究所等单位合作，以多专业协同、多维度思考的设计方式，经济高效的利用土地资源，兼顾当下与未来的可持续发展。

立体山地海绵系统引导新城开发

后河环境综合整治工程是其所在片区的先导性项目，先于周边规划地块建设。我们综合考虑了项目所在地现存或将要面临的问题：周边城区的开发建设可能对现状自然河谷造成环境破坏，自然河谷难以应对未来城市的污染压力，土地综合利用有待提高等。

我们提出以立体山地海绵系统为基础，探索具有示范意义的CBA（clean、beautiful、active）新城开发模式，以洁净、美丽、活力为目标，将海绵涵养廊道、生态景观廊道、活力休闲廊道合一，切实保护自然河谷生态本底，智慧应对新城开发。让自然做功，自然积

区域鸟瞰效果图
A regional bird's-eye view rendering

存、自然渗透、自然净化，综合应用“渗、滞、蓄、净、用、排”措施，流域综合整治、天然海绵保护修复、生态景观营造相互融合，展现青山绿水自然景观，构建新城生态安全屏障。设计兼顾当下与未来，适应城市开发，分期建设：近期完成流域水环境整治、自然山地海绵体的保护提升，基本建成沿河绿道；远期完善绿道服务设施，发挥综合效益，提升土地功能价值。

保护山谷现有植被体系，恢复山地生态系统，充分发挥植被对雨水的滞留作用，降低年雨量径流系数。改造现有梯田，使其变为具有蓄滞净化作用的堰塘及净水梯田。针对山地系统表层土薄、土壤较贫瘠、渗透性较差等特点，设置植草沟将雨水导引至堰塘及净水梯田，有效实现初期雨水净化。将原有化学试剂厂改造为旁路式生态塘，发挥净水作用，兼作水体调蓄。调蓄雨水用于绿化灌溉、河道补水等。充分利用现状水利设施蓄水，就近补给净水梯田和生态塘，保持LID（low impact development，低影响开发）系统在旱季时仍可运行。利用现状冲沟设置超标雨水排放系统，防止对山体的冲刷和对LID系统的破坏。

建立智能绿道系统，引领绿色生活

绿道设计总长23.9千米，包含主线和支线两个等级，山地、滨水两个类型。将绿道作为功能复合的网络结构，发挥交通互联、休闲健身、生态保护等社会、文化、旅游、经济多方面的积极作用。绿道游径布局因山就势，对现状道路进行补充与完善，预留与规划路网的衔接点，合理组织动静态交通。营造多层级的户外休闲空间，引入智能健身系统，引导科学健身。结合绿道主题活动策划，展示山地海绵的相关知识，开发户外教育课程，以轻资产运营，促进文旅休闲产业发展。

山地海绵 创新示范

流域整治、山地海绵、生态景观工程一体化设计，让河道风景融入城市生活

山地海绵剖面效果图
Mountain sponge profile rendering

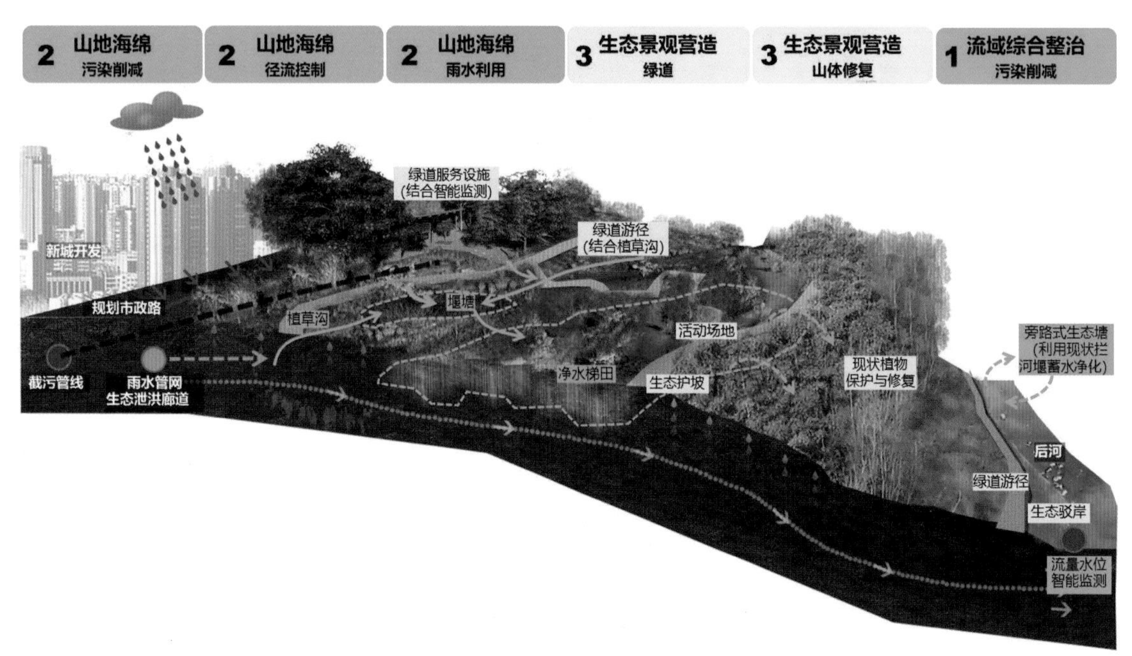

立体山地海绵设计思路
Design concept for three-dimensional mountain sponge system

在立体海绵系统的基础上，营造多层级的开放空间，依托智慧网络加强体验性，激发场所活力
Creating multi-level open spaces based on the three-dimensional sponge system, enhancing the experiential aspect through smart networks and activating the vitality of the place

结合现状梯田布局山城特色立体海绵系统
Three-dimensional sponge system combined with the existing terraced layout, highlighting the characteristics of a mountain city

水体常水位与高水位景观效果图
Landscape rendering showing the water level at normal and high levels

建成实景照片 Real-life photos of the completed project

建成实景照片　Real-life photos of the completed project

02 湖南益阳市梓山湖片区规划设计

Hunan Yiyang City Zishan Lake Scenic Area (park) Urban Design

- **项目地点：** 中国 益阳
- **项目规模：** 7.13 平方千米
- **设计时间：** 2016-2017 年
- **合作单位：** 中国建筑设计研究院崔愷工作室

- **Project location:** Chongqing, China
- **Project scale:** 7.13 square kilometers
- **Design period:** 2016-2017
- **Cooperating unit:** Cui Kai Studio of China Architecture Design & Research Group

梓山湖是湖南省益阳市主城区最大的湖泊，我们以梓山湖水质保护为中心，采取综合治理措施，实现梓山湖区域排涝安全，实现人与自然的和谐发展。

Zishan Lake is the largest lake in the central urban area of Yiyang City, Hunan Province. We prioritize the protection of water quality in Zishan Lake and adopt comprehensive management measures to ensure safe water-logging drainage in the Zishan Lake area. Our aim is to achieve harmonious development between humans and nature, as well as between humans and water.

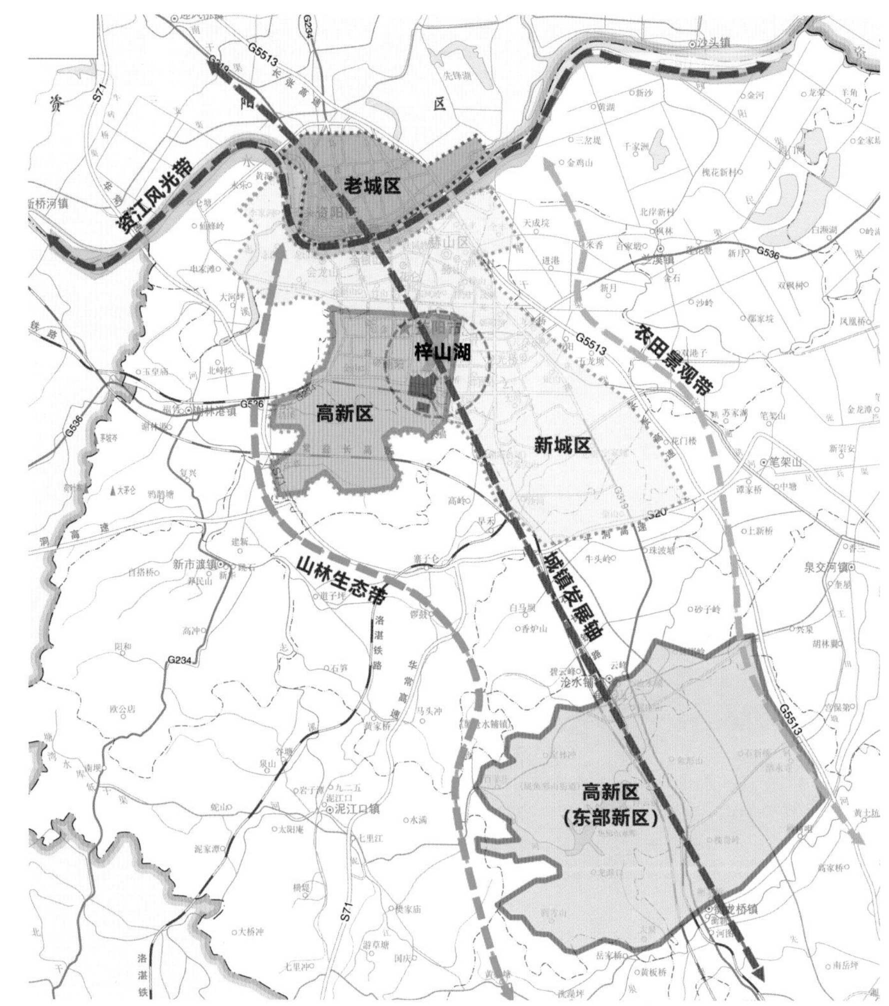

梓山湖区位图

Zishan Lake location map

湖南省益阳市位于长江中下游平原的洞庭湖南岸，是环洞庭湖生态经济圈的核心城市之一，也是长株潭3+5城市群之一。城市得名于益水，自古是江南富饶的“鱼米之乡”。据考证古益水即为资江，益阳老城原位于资江北岸，随着城市进一步向东南跨江拓展，梓山湖成为了益阳主城区的地理中心。

梓山湖原是以汇水灌溉功能为主的小型水库，水域面积约1平方千米，是益阳主城区最大的湖泊。梓山湖岸线曲折，湖畔丘陵起伏，周边山体植被覆盖率大于80%，山谷中散布有少量现状村落。梓山湖虽然是益阳市重要的生态绿核，但湖区已逐渐受到城市建设的侵蚀。目前梓山湖北岸高楼林立，西部为高尔夫球场及奥林匹克公园，南部拟建益阳文化政务中心，西北、西南、东北、东南四面均为已建成或规划中的居住及商业用地。项目用地位于湖区东部，面积约2.45平方千米。

中国建筑设计研究院崔愷工作室承接了梓山湖南岸的益阳文化政务中心建筑设计工作，我们与之合作开展梓山湖片区城市设计、梓山湖公园修建性详细规划及旅游项目策划等工作。

生态优先，构建山湖海绵体

梓山湖属于城市人工湖，没有其他湖泊或河流接入，湖水由雨水汇集而成。近年来由于降雨量减少、过度开发、市政设施不到位、集雨面积减少等原因，梓山湖库容减少、水位下降、水质状况走低，不仅难以保证农田灌溉的需要，整体的生态功能也在退化。因此，对于梓山湖片区的规划设计以梓山湖水质保护为中心，采取综合治理措施，实现人与自然、人与水的和谐发展，实现梓山湖区域排涝安全，实践梓山湖海绵城市建设理念。

通过补水系统设计和现状溢流堰、防洪闸系统，控制湖水水量收支平衡，保证湖水水位在合理范围。采用生态净化塘、人工湿地等手段，最大限度地模拟水体自净过程，以较高的效率实现微污染水处理；周边绿化带建设生物滞留带，净化周边绿地雨水径流，控制面源污染；沿景观驳岸引入水生植被，结合现有湖内水生动物，建立完善的生态系统，并将生态系统控制在较为理想的状态，削减内源污染。建立物联网监测控制系统平台，深层次获取各项监测指标，为梓山湖水质、水量保持提供数据基础。针对暴雨、连续干旱、突发性水质事件等，建立相应的工程措施与管理制度。

统筹联动，打造城市活力中心

鉴于梓山湖融山、水、林于一体的资源优势和优越区位条件，该片区兼具城市公园与旅游景区的双重属性。我们从统筹协调的角度出发，结合已出让地块等规划条件，合理划定片区的绿线范围，有效保护梓山湖及周边生态环境空间，避免城市建设的进一步侵蚀。

梓山湖片区城市设计以“两轴分主题，一环串五区”的布局结构展开，致力于将该区域从城市的地理中心转变为承载市民生活、展示城市形象的新兴活力中心。其中横轴联系梓山湖公园与高尔夫球场、奥林匹克公园，三者互为补充，突出健康活力主题；纵轴联系北部市民广场与南部文化政务中心，突出城市文化主题。“五区”分别为“北商、南文、东游、西体、中心湖区”，由全长7500米的环园路串联，形成一个功能复合的整体。

益阳是著名的“羽毛球之乡”，梓山湖西部的奥林匹克公园是中国羽毛球队训练基地，该片区也是主要的市民体育活动场所，已有不少市民自发进行徒步、骑行等健身活动。我们希望能够传承市民的活力基因，联动周边现有体育运动场所，创建湖南省首个国家级全民健身户外活动基地，打造全年龄的生态运动休闲公园。

梓山湖南部拟建益阳文化政务中心建筑群，包含规划馆、博物馆、群艺馆、图书馆、市民活动中心、政务中心等，未来将成为益阳新地标，是城市文化会展、市民公共活动的重要场所。公园与公共建筑群同步进行规划设计，达到二者的互动与交融，共同打造山清水秀、有文化活力的城市中心风景区。

总平面图
Master plan

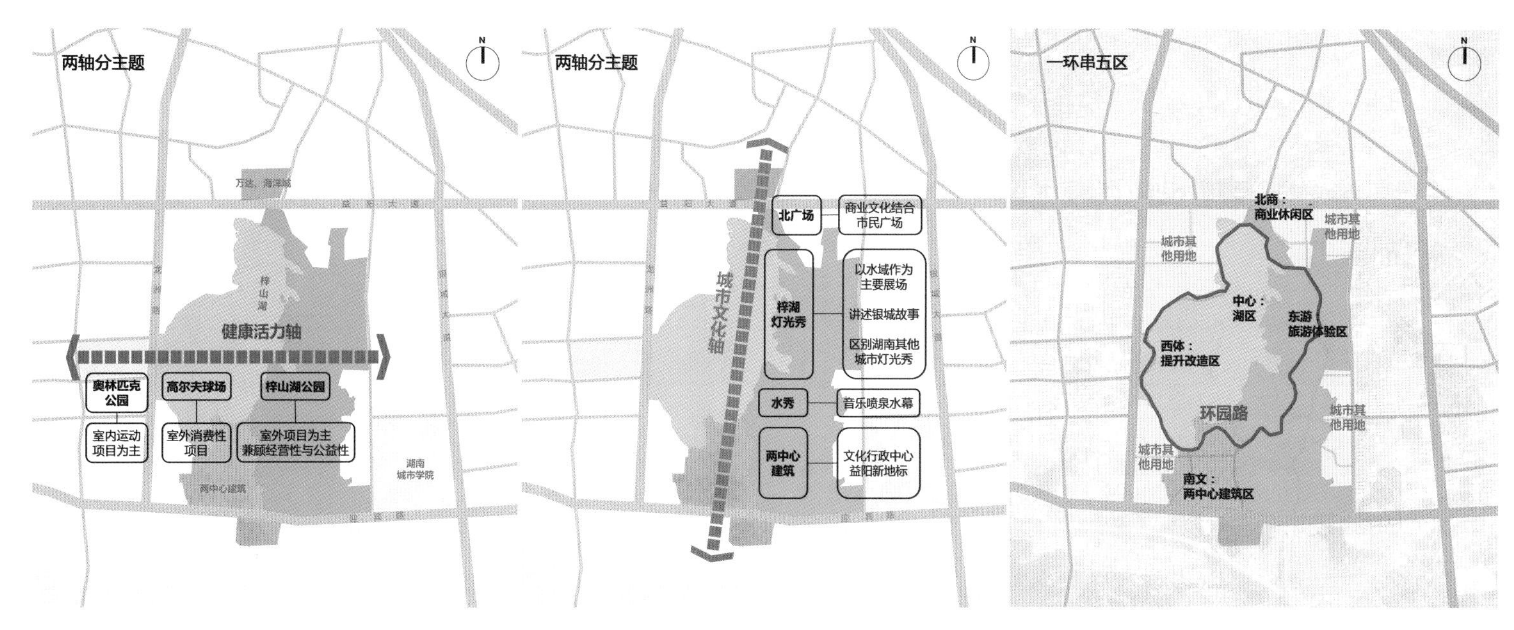

布局结构　Layout structure

鸟瞰效果图　A bird's-eye view rendering

水位下降，消落带淤泥堆积

雨污合流直排入湖

局部水体浑浊、黑臭

原有场地现状　The status quo of the existing site

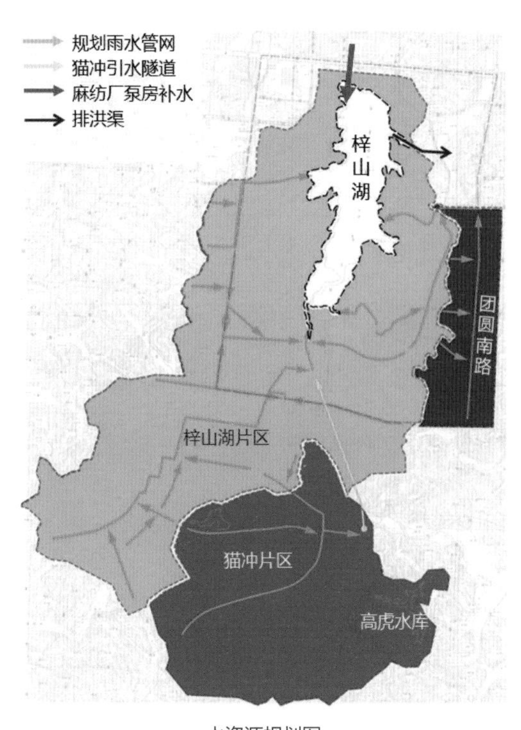

水资源规划图

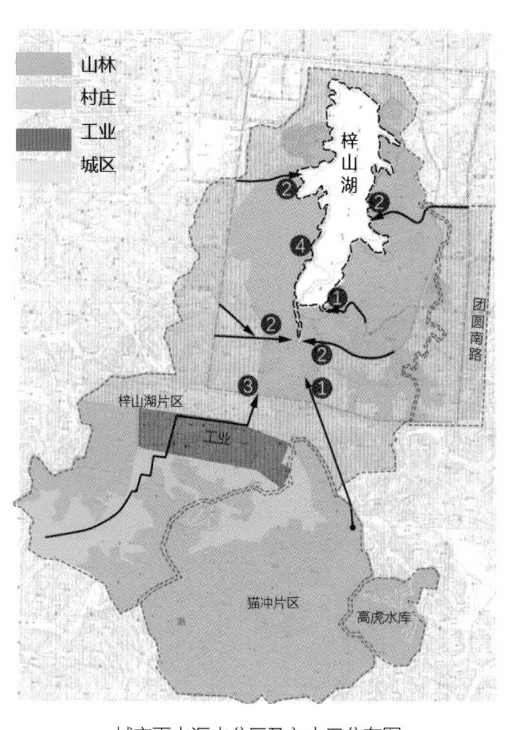

城市雨水汇水分区及入水口分布图

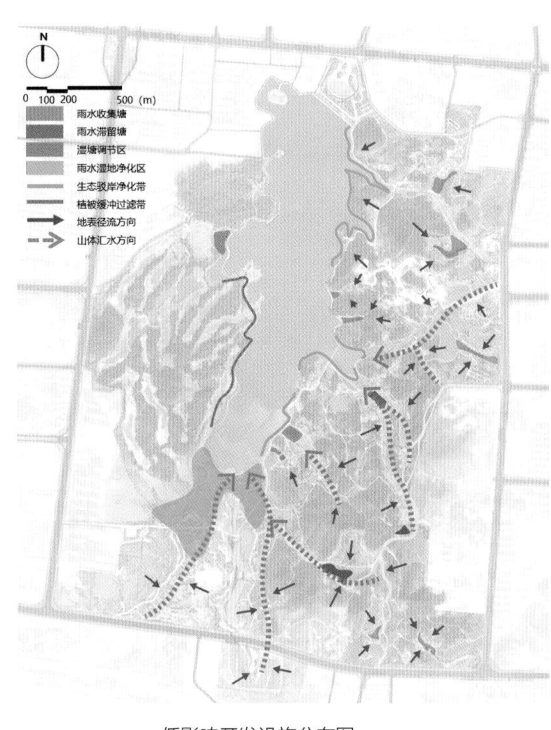

低影响开发设施分布图

梓山湖低影响开发系统　Low-impact development system for Zishan Lake

城市汇水处理流程

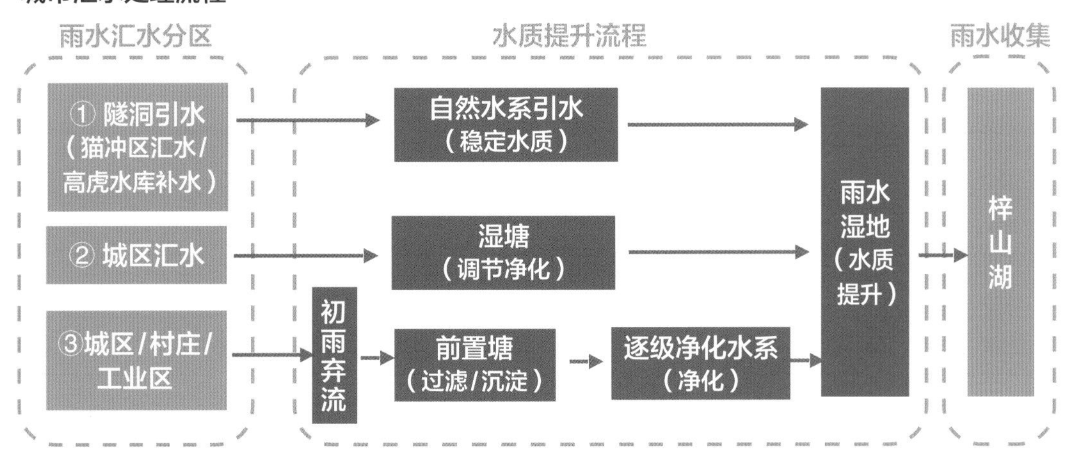

公园内雨水收集流程

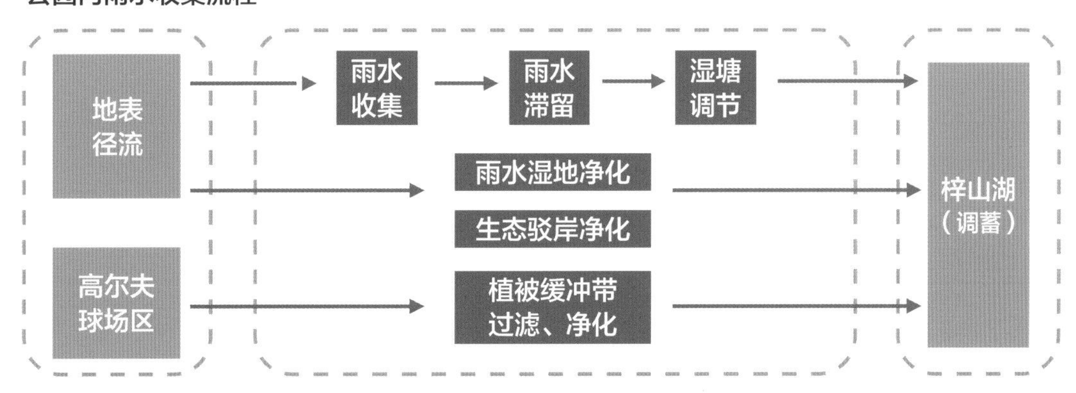

雨水湿地结合科普展示

雨水汇水系结合活动空间

梓山湖低影响开发系统　Low-impact development system for Zishan Lake

环湖绿道效果图　Greenway along the lake effect rendering

健身环路效果图　Fitness loop trail effect rendering

花溪景观带效果图　Flower creek landscape belt effect rendering

竹林文艺带效果图　Bamboo forest cultural belt effect rendering

擂茶休闲带效果图　Ground Tea leisure belt effect rendering

黑茶养生带效果图　Dark Tea health belt effect rendering

后　　记

无界景观工作室成立于2004年，一直在城市更新、乡村振兴、城乡融合、生态修复、文旅发展、历史文化保护与传承等领域，致力于探索景观设计融合多学科智慧、统筹多专业协作的最行之有效的途径与方法，助力社会、经济和生态的绿色可持续发展。

工作室始终关注人的日常生活与公共空间的联系，关注景观设计对社会关系的良性引导，对生态系统的保护修复，探寻因地、因时、因人而异的，与同质化相悖的解决方案；始终坚持以专业手段协调人与环境的关系，将风景融入日常生活，缓解人的生存压力，激发民众活力与场地生产力，提升“安住”者的幸福感与归属感。

格物而致知，笃行而致远。未来，愿行而不辍，继续为城、为乡、为人、为生境，营建整体的、连续的美丽。

AFTERWORD

Founded in 2004, View Unlimited Landscape Architecture Studio has been dedicated to exploring the most effective ways and methods for integrating multidisciplinary ideas and coordinating multi-disciplinary collaboration in the fields of urban renewal, rural revitalization, urban-rural integration, ecological restoration, cultural tourism and development, and historical and cultural preservation and inheritance, to support the green and sustainable development of our society, economy and ecology.

We focus on the inherent connection between people's daily life and public spaces, with an emphasis on the constructive integration of landscape design into social relationships, the protection and restoration of the ecosystem, and the search for solutions that veer away from homogenization by taking into account the surrounding environment, time period, and community. We strive to harmonize the relationship between people and the environment by employing professional techniques and integrating landscapes into everyday life that relieve the pressures of human existence, stimulate people's vitality and productivity, and ultimately enhance a sense of happiness and belonging in those who simply wish to “abide”.

Attaining knowledge from our origins, our actions echo far beyond. Our aim is to persevere in creating a holistic and sustainable aesthetic for cities, towns, people, and our natural habitat.

致谢

感谢本书所有项目中合作过的业主、建筑师、规划师及众多相关参与者，没有你们就没有今天呈现在大家面前的全面成果！感谢集团和设计院领导对工作室多年来毫无保留的支持和爱护，以及同仁、朋友们给予的支持与鼓励。最后，还要感谢自工作室成立以来共同奋斗过的每一位同事，感谢你们的陪伴和在项目中付出的努力！